Materials, Productions, Exchange Network and their Impact on the Societies of Neolithic Europe

Proceedings of the XVII UISPP World Congress
(1–7 September 2014, Burgos, Spain)

Volume 13/Session A25a

Edited by

Marie Besse and Jean Guilaine

Archaeopress Archaeology

ARCHAEOPRESS PUBLISHING LTD
Gordon House
276 Banbury Road
Oxford OX2 7ED

www.archaeopress.com

ISBN 978 1 78491 524 7
ISBN 978 1 78491 525 4 (e-Pdf)

© Archaeopress, UISPP and authors 2017

VOLUME EDITORS: Marie Besse and Jean Guilaine

SERIES EDITOR: The board of UISPP

CO-EDITORS – Laboratory of Prehistoric Archaeology and Anthropology, Department F.-A. Forel for Environmental and Aquatic Sciences, University of Geneva

SERIES PROPERTY: UISPP – International Union of Prehistoric and Protohistoric Sciences

Proceedings of the XVII World UISPP Congress, Burgos (Spain)
September 1st - 7th 2014

KEY-WORDS IN THIS VOLUME: Neolithic, Europe, Materials, Productions, Exchange Networks

UISPP PROCEEDINGS SERIES is a printed on demand and an open access publication, edited by UISPP through Archaeopress

BOARD OF UISPP: Jean Bourgeois (President), Luiz Oosterbeek (Secretary-General), François Djindjian (Treasurer), Ya-Mei Hou (Vice President), Marta Arzarello (Deputy Secretary-General). The Executive Committee of UISPP also includes the Presidents of all the international scientific commissions (www.uispp.org)

BOARD OF THE XVII WORLD CONGRESS OF UISPP: Eudald Carbonell (Secretary-General), Robert Sala I Ramos, Jose Maria Rodriguez Ponga (Deputy Secretary-Generals)

All rights reserved. No part of this book may be reproduced, or transmitted, in any form or by any means, electronic, mechanical, photocopying or otherwise, without the prior written permission of the copyright owners.

This book is available direct from Archaeopress or from our website www.archaeopress.com

Contents

List of Figures and Tables ii

Foreword to the XVII UISPP Congress Proceedings Series Edition iv
Luiz Oosterbeek

Foreword v
Jean Guilaine and Marie Besse

White-painted Pottery in the Early Neolithic Balkans 1
Darko Stojanovski

**Settlements – Head and Settlements – Tail in the Neolithic Obsidian Exchange Network
in the Western Mediterranean** 13
Tania Quero

**Original and Skeuomorph: On the materiality of the Chalcolithic package of prestige
in South Eastern Europe** 17
Dragoş Gheorghiu

**Exchange and interaction: the Iberian Mediterranean between
the VI and III millennia cal BC** 27
Teresa Orozco Köhler and Joan Bernabeu Aubán

**The Western network revisited: the transition into agro-pastoralism in
the Alto Ribatejo, Portugal** 39
Nelson J. Almeida, Cristiana Ferreira, Sara Garcês, Ana Cruz, Pierluigi Rosina and
Luiz Oosterbeek

Mobility in late Prehistory in Galicia: a preliminary interpretation from pottery 51
M Pilar Prieto Martínez and Óscar Lantes Suárez

**Types and gesture. The jewellery of the Copper age in the Alps in
a techno-typological study** 69
Stefano Viola, Maria Adelaide Bernabo' Brea, Dino Delcaro, Federica Gonzato,
Cristina Longhi, Giorgio Gaj, Roberto Macellari, Luciano Salzani,
Alessandra Serges, Iames Tirabassi, Marie Besse

List of Figures and Tables

D. Stojanovski: **White-painted Pottery in the Early Neolithic Balkans**

FIGURE 1. WHITE-PAINTED POTTERY (WPP) GROUPS IN THE BALKAN PENINSULA 3
FIGURE 2. CALIBRATION PLOT FOR NEA NIKOMEDEIA (GROUP I) 4
FIGURE 3. CALIBRATION PLOT FOR GROUP II 4
FIGURE 4. CALIBRATION PLOT FOR GROUP III 5
FIGURE 5. CALIBRATION PLOT FOR GROUP IV 6
FIGURE 6. CALIBRATION PLOT FOR GROUP V 7
FIGURE 7. CALIBRATION PLOT FOR KARANOVO II (GROUP VI) 8

T. Quero: **Settlements – Head and Settlements – Tail in the Neolithic Obsidian Exchange Network in the Western Mediterranean**

FIGURE 1. SITES POSITION: S. MARTINO SPADAFORA AND GRANAROLO DELL'EMILIA 14
FIGURE 2. S. MARTINO SPADAFORA SITE: TECHNOLOGICAL DISTRIBUTION OF
   THE OBSIDIAN KNAPPING PRODUCTS 14
FIGURE 3. S. MARTINO SPADAFORA SITE: OBSIDIAN CORES AND PRODUCTS. TOOLS ARE SCRAPERS,
   DENTICULATES AND PERFORATORS 14
FIGURE 4. GRANAROLO SITE: RAW MATERIALS DISTRIBUTION AND THE KNAPPING PRODUCTS.
   THE BLADELETS ARE UNRETOUCHED 15

D. Gheorghiu: **Original and Skeuomorph: On the materiality of the Chalcolithic package of prestige in South Eastern Europe**

FIGURE 1. GRAVE 4 (CENOTAPH), VARNA CEMETERY (VARNA MUSEUM) 18
FIGURE 2. GRAVE 43 (VARNA MUSEUM) 19
FIGURE 3. GRAVE 43, DETAIL. (VARNA MUSEUM) 20
FIGURE 4. A REPLICA OF THE SULTANA PACKAGE (OLTENIȚA MUSEUM) 21
FIGURE 6. THE HORODNICA II PACKAGE (NATURAL HISTORY MUSEUM, VIENNA) 22
FIGURE 5. THE BRAD PACKAGE (ROMAN MUSEUM OF HISTORY) 22

T. Orozco Köhler and J. Bernabeu Aubán: **Exchange and interaction: the Iberian Mediterranean between the VI and III millennia cal BC**

FIGURE 1. SOME OF THE RAW MATERIAL SOURCES RECENTLY DISCOVERED IN IBERIA:
   PICO CENTENO (VARISCITE), CASA MONTERO (FLINT) 29
FIGURE 2. AREA OF SCHIST BRACELETES DISTRIBUTION IN THE MEDITERRANEAN IBERIA
   ALONG THE EARLY NEOLITHIC 30
FIGURE 3. PIECES SHOWING – THE MANUFACTURING PROCESS OF THE SCHIST BRACELETS AND
   THE REUSE OF THESE OBJECTS AS PENDANTS 31
FIGURE 4. THE DISTRIBUTION OF AMPHIBOLITE STONE AXES FROM THE SOURCE AREA IN
   THE SOUTHEAST ARRIVES TO THE CENTRAL MEDITERRANEAN 32
FIGURE 5. SOME TECHNOLOGICAL FEATURES ALLOW TO DELIMIT CLEARLY REGIONAL DIFFERENCES IN IBERIA 32
FIGURE 6. HORNFELS AXES AND ADZES HAVE THEIR SOURCE AREA IN THE NORTHEAST MEDITERRANEAN 33
FIGURE 7. THE EYE-MOTIF IS ADOPTED AND ADAPTED TO A VARIETY OF SUPPORTS OVER
   A WIDE GEOGRAPHICAL FRAMEWORK, WITH REGIONAL DIFFERENCES 34

N. J. Almeida: **The Western network revisited: the transition into agro-pastoralism in the Alto Ribatejo, Portugal**

FIGURE 1. DIGITAL ELEVATION MODEL OF THE ALTO RIBATEJO WITH LOCATION OF THE SITES 40
TABLE 1. DATINGS FOR THE MAIN CONTEXTS REFERRED IN THE TEXT 41
TABLE 2. GENERAL ABUNDANCE OF THE MAIN TAXA IDENTIFIED IN THE ALTO RIBATEJO PALYNOLOGICAL(^)
   AND ANTHRACOLOGICAL RECORDS 43
TABLE 3. GENERAL ABUNDANCE OF THE MAIN TAXA IDENTIFIED IN THE ALTO RIBATEJO ZOOARCHAEOLOGICAL
   RECORDS BASED ON NISP VALUES 44

M. P. Prieto Martínez and Ó. Lantes Suárez: **Mobility in late Prehistory in Galicia: a preliminary interpretation from pottery**

Figure 1. Pottery analysed from the Early Neolithic. 1. Map showing the distribution of the Early Neolithic sites in Galicia, 2. Fragments of the same vessel from the site of As Mamelas with boquique decoration, analysed in this study. 3. Distribution of areas with boquique decoration from the Neolithic period in the Iberian Peninsula 53

Figure 2. Pottery analysed from the Mid-Neolithic. 1. Map showing the distribution of Mid-Neolithic sites in Galicia. 2. Fragments of the vessels analysed 55

Figure 3. Pottery analysed from the Late Neolithic. 1. Map showing the distribution of Late Neolithic sites in Galicia. 2. Map showing the distribution of the 3 pottery areas of the Iberian Peninsula. 3. Selection of some of the vessels analysed 57

Figure 4. Pottery analysed from the Early Bronze Age. 1. Distribution of regional Bell Beaker styles in the Iberian Peninsula. 2. Map showing the distribution of Early Bronze Age sites in Galicia. 3. Selection of some of the vessels analysed. 4. Interpretation of the possible long and short distance relationships at this time 59

Figure 5. Pottery analysed from the Late Bronze Age. 1. Map showing the distribution of Late Bronze Age sites. 2. Selection of some of the vessels that were analysed. 3. Distribution of sites with WHR pottery in the NW Iberian Peninsula and interpretation of the possible relationships with other parts of Europe through the stamped decoration on the WHR pottery 63

Table 1a. Minerology of the pottery analysed from the Early Neolithic 54
Table 1b. Minimum distances from the site to the most likely sources of raw material for the pottery from the Early Neolithic 54
Table 2a. Mineralogy of the pottery analysed from the Mid-Neolithic 56
Table 2b. Minimum distances from the site to the most likely sources of raw material for the pottery from the Mid-Neolithic 56
Table 3a. Mineralogy of the pottery analysed from the Late Neolithic 58
Table 3b. Minimum distances from the site to the most likely sources of raw material for the pottery from the Late Neolithic 58
Table 4a. Mineralogy of the pottery analysed from the Early Bronze Age 60
Table 4b. Minimum distances from the site to most likely sources of raw material for the pottery from the Early Bronze Age 60
Table 5a. Mineralogy of the pottery analysed from the Late Bronze Age 62
Table 5b. Minimum distances from the site to most likely sources of raw material for the pottery from the Late Bronze Age 62

S. Viola et al.: **Types and gesture. The jewellery of the Copper age in the Alps in a techno-typological study**

Figure 1. Localization of the sites studied and hole making tolls before and after use 72
Figure 2. Manufacture sequence of Copper Age, disc and cylindrical beads-medium size 73
Figure 3. Manufacture sequence of Copper Age, disc and cylindrical beads-small size 74
Figure 4. Manufacture sequence of Copper Age, long beads 75
Figure 5. Manufacture sequence of Bell Beaker Culture, biconical and globular beads 76
Figure 6. Manufacture sequence of Early Bronze Age, discoids, cylindrical beads-medium size 77
Table 1. The list of sites 70
Table 2. The experimental tests: aspects, variables and descriptions 71
Table 3. The perforation tests 71

Foreword to the XVII UISPP Congress Proceedings Series Edition

Luiz Oosterbeek
Secretary-General

UISPP has a long history, starting with the old International Association of Anthropology and Archaeology, back in 1865, until the foundation of UISPP itself in Bern, in 1931, and its growing relevance after WWII, from the 1950's. We also became members of the International Council of Philosophy and Human Sciences, associate of UNESCO, in 1955.

In its XIVth world congress in 2001, in Liège, UISPP started a reorganization process that was deepened in the congresses of Lisbon (2006) and Florianópolis (2011), leading to its current structure, solidly anchored in more than twenty-five international scientific commissions, each coordinating a major cluster of research within six major chapters: Historiography, methods and theories; Culture, economy and environments; Archaeology of specific environments; Art and culture; Technology and economy; Archaeology and societies.

The XVIIth world congress of 2014, in Burgos, with the strong support of Fundación Atapuerca and other institutions, involved over 1700 papers from almost 60 countries of all continents. The proceedings, edited in this series but also as special issues of specialized scientific journals, will remain as the most important outcome of the congress.

Research faces growing threats all over the planet, due to lack of funding, repressive behavior and other constraints. UISPP moves ahead in this context with a strictly scientific programme, focused on the origins and evolution of humans, without conceding any room to short term agendas that are not root in the interest of knowledge.

In the long run, which is the terrain of knowledge and science, not much will remain from the contextual political constraints, as severe or dramatic as they may be, but the new advances into understanding the human past and its cultural diversity will last, this being a relevant contribution for contemporary and future societies.

This is what UISPP is for, and this is also why we are currently engaged in contributing for the relaunching of Human Sciences in their relations with social and natural sciences, namely collaborating with the International Year of Global Understanding, in 2016, and with the World Conference of the Humanities, in 2017.

The next congress of UISPP, in Paris (2018), will confirm this route.

Foreword

Jean Guilaine
Collège de France

Marie Besse
Université de Genève, Department F.-A. Forel for Environmental and Aquatic Sciences,
Laboratory of Prehistoric Archaeology and Anthropology

Scholars who will study the historiography of the European Neolithic, more particularly with regards to the second half of the 20th century and the beginning of the 21st century, will observe a progressive change in the core understanding of this period. For several decades the concept of 'culture' has been privileged and the adopted approach aimed to highlight the most significant markers likely to emphasise the character of a given culture and to stress its specificities, the foundations of its identity. In short, earlier research aimed primarily to highlight the differences between cultures by stressing the most distinctive features of each of them. The tendency was to differentiate, single out, and identify cultural boundaries. However, over the last few years this perspective has been universally challenged. Although regional originality and particularisms are still a focus of study, the research community is now interested in widely diffused markers, in medium-scale or large-scale circulation, and in interactions that make it possible to go beyond the traditional notion of 'archaeological culture'. The networks related to raw materials or finished products are currently leading us to re-think the history of Neolithic populations on a more general and more global scale. The aim is no longer to stress differences, but on the contrary to identify what links cultures together, what reaches beyond regionalism in order to try to uncover the underlying transcultural phenomena. From culturalism, we have moved on to its deconstruction. This is indeed a complete change in perspective. This new approach certainly owes a great deal to all kinds of methods, petrographic, metal, chemical and other analyses, combined with effective tools such as the GIS systems that provide a more accurate picture of the sources, exchanges or relays used by these groups. It is also true that behind the facts observed there are social organisations involving prospectors, extractors, craftsmen, distributors, sponsors, users, and recyclers. We therefore found it appropriate to organise a session on the theme 'Materials, productions, exchange networks and their impact on the societies of Neolithic Europe'.

How is it possible to identify the circulation of materials or of finished objects in Neolithic Europe, as well as the social networks involved? Several approaches exist for the researcher, and the present volume provides some examples.

Let us take the case of the white painted ware in the Early Neolithic of the Balkan Peninsula.

D. Stojanovski shows how several cultural groups exhibited this particular pottery decoration. However, according to D. Stojanovski, it is not very likely that such a technique would have been transmitted by migrants who carried with them the 'Neolithic package'. In this particular case we are instead dealing with a borrowing process adopted (or not) in various ways according to distinct factors: the attractiveness of its visual effect, traditions, needs, environmental context.

It is a known fact that obsidian is a perfect marker of exchange relationships across the Mediterranean basin throughout the Neolithic. T. Quero however points to differences with regards to these transfers, based on two examples in Italy. The first site, in Northern Sicily, near the lava flow of Lipari, is a place of obsidian redistribution throughout the Neolithic. This material is abundantly used, although the characteristics of production change over time: over-exploited nuclei generating mainly a flake industry during the Stentinello, rise of blade production by pressure during the Diana stage. At the second site, in Northern Italy, far away from the sources of supply, the imported material is rare and has only exotic value.

D. Gheorghiu develops an original example stemming from the Chalcolithic period in South-East Europe. Here the elites display wealth by using valorised objects of allochthonous origin. The wealthy dead of the Varna cemetery, for example, are distinguished by an authentic 'package' of objects revealing a real ideology of prestige. This model, conveying both concepts of ideas and technical transfers, will be exported again towards peripheral cultures such as the Tripolje culture. When groups that settled at the margins were not able to acquire pieces from the source, they replace these original pieces with similar creations (skeuomorphs) that make it possible, through its productions, to maintain a hierarchic social model.

T. Orozco Köhler and L. Bernabeu Aubán provide some examples of circulation networks during the Neolithic on the Iberian Peninsula. They first emphasise the progress achieved with regard to the identification of sources of supply: flint mines of de Casa Montero (Madrid), open-air exploitations of Andalusian obsidian at Pico Centero (Huelva). Several axes of circulation are mentioned: shale bracelets in the south, amphibolite axes around Valencia, Sardinian obsidian brought to Catalonia. The authors highlight the variations affecting these networks in space and time. They also emphasise their possible superimposition. They demonstrate that these circulations were not restricted to the dissemination of objects but involved displacements of individuals within a context governed by social motivations (alliances, relationships between individuals).

The establishment of an agro-pastoral economy in the Alto Ribatejo, in the Centro region of Portugal, was studied by N. J. Almeida, C. Ferreira, S. Garcês, A. Cruz, P. Rosina, and L. Oosterbeek. The archaeological data demonstrates a great variety of situations in the distribution of the sites, depending on space and time. It is thought that this research can be improved by applying new methods likely to better reflect the reality of settlements throughout the Neolithic. After reviewing the evolution of the environmental setting under the effect of human pressure (opening of the landscape, fauna), the authors point to the impact of coastal arrivals, vehicles of the neolithisation dynamic, but also, in parallel, to the role of more continental areas which have their own specificity in the emergence of cultural productions.

M. P. Prieto Martinez and O. Lantes Suárez look at the example of pottery as an element which makes it possible to assess the existence of circulation networks in Galicia from the Neolithic to the Bronze Age. Based on the typology and on archaeometric analyses, these authors attempt to estimate the distance that separates sites of discovery from the elaboration space of the ceramics. This method proposes an alternative for the simple local/non-local alternative. The authors propose five models for this grid (ranging between 0 and more than 200 km), with the district scale (between 7 and 50 km) being the most frequent. This approach also provides a more appropriate picture of the true mobility of the groups (for example possible settlement instability during the Middle Neolithic).

An alternative way of analysing mobility is based on the technical analysis of the ornaments. The example cited by S. Viola, M. A. Bernabò Brea, D. Delcaro, F. Gonzato, C. Longhi, G. Gaj, R. Macellari, L. Salzani, A. Serges, I. Tirabassi, and M. Besse, refers to the stone ornaments of the Alpine Chalcolithic. The operational sequence of manufacturing, distinct technical details, and even use wear are all specific markers of an object during its displacements. These remains are also identifiers of the definition of territories and of cultural groups.

By giving preference to various aspects of raw material or finished product analyses, it is obvious how current research makes it possible to draw a more accurate and a more complex picture of the Neolithic circulation networks, as well as simultaneously producing a more balanced one of the distances covered at the time. The examples cited in this volume confirm that the first agricultural communities, through the establishment of networks of varying and sometimes contrasting scales, were authentic 'exchange-based societies'.

White-painted Pottery in the Early Neolithic Balkans

Darko Stojanovski
Department of Geology, Universidade de Trás-os-Montes e Alto Douro,
Vila Real, Portugal; IDQP (stjdrk@unife.it)

Abstract
One of the most emblematic and most (ab)used feature of the early Neolithic in the central and northern Balkans is the White-painted Pottery (WPP). This article attempts to bring a somewhat clearer presentation of the distribution of this pottery style in time and territory. Through the emerging patterns, questions concerning the mechanisms behind the neolithisation process in the Balkans are addressed.
Key words: *White-painted Pottery (WPP), neolithisation, chronology, Balkans*

Résumé
Une des caractéristiques les plus emblématiques du Néolithique ancien dans les Balkans du centre et du nord est le White-painted Pottery (WPP). Cet article tente d'apporter une présentation un peu plus claire de la répartition de ce style de poterie dans le temps et dans le territoire. Selon les tendances émergentes, nous discuterons les questions concernant les mécanismes à l'origine du processus de néolithisation dans les Balkans.
Mots clés: *White-painted Pottery (WPP), néolithisation, chronologie, Balkans*

Introduction

The history of pottery appearance and further development is still controversial and widely discussed in the Balkan Peninsula. The major issues concerning pottery are firmly connected with the beginning of farming, the appearance of ground-stone items, and the first permanent settlements – all elements of a new socio – technological stage in human development. An overgeneralization of these relations during the past century existed in the form of the 'Neolithic package' concept, postulating a uniform material culture – a set of elements appearing together over a wide territory, including Southwest Asia, Anatolia and the Balkans. This concept also implies migration of people as a mechanism for diffusion of these objects and the technology to produce them (Perlès, 2001a, 2001b; Özdoğan, 2010, 2011). With the introduction of new, more eclectic models for neolithisation, which are slowly surpassing the migration vs. indigenous duality, the 'Neolithic package' concept started to lose its relevance (Çilingiroglu, 2005; Reingruber, 2011). Under this layer of general uniformity or similarity, there is a great variety in any aspect of the elements of this package. This is especially evident in the most common element – the pottery.

The initial stages of pottery production in the Balkans show great variability from one region to another. In some, for example Macedonia, Southeast Albania, and East Croatia, the white-painted pottery (hereafter WPP) is present since the beginning of the Neolithic, together with what in an isolated context would be called 'monochrome pottery' (Minichreiter, 2007; Naumov et al., 2009; Bunguri, 2014). In other regions, like Southwest and North Bulgaria and Serbia, some authors still maintain the idea of a 'monochrome phase' as the initial stage of pottery development, followed by the white-painted pottery phase (Тодорова and Вайсов, 1993; Чохаджиев, 2007; Bogdanovich, 2007). However, except for Krainici (Чохаджиев et al., 2007), there is very little secure evidence of such diachronic relationship. The improbability of existence of a separate chronological phase with only 'monochrome' ware has been elaborated in details elsewhere (Krauß, 2011). To this we can add the example from Eastern Macedonia, where a settlement called Grncharica, containing only 'monochrome' pottery was discovered. A burial from the same short-lived site was dated to the 56th – 57th century cal BC, i.e. following what would be considered as 'WPP phase' (Stojanovski et al., 2014), thus, in this mode of reasoning, the site is displaying a reverse diachronic order.

Clearly the appearance of pottery in these territories is much more complicated than a picture where two pottery modes are representing two successive sequences in time. Trying to comprehend the mechanisms behind the introduction of this technology throughout the peninsula, with preconceptions embedded in theories of a uniform material culture being brought by large migrating groups of people, is one of the reasons for the controversial nature of this subject.

Even though rare, the WPP is a constant occurrence in the early Neolithic sites in the Balkans. The visual flamboyancy has made it one of the most famous features of the period, which often led to overemphasising its role and meaning. This is usually a very small part of an assemblage, occurring at different points in time at different regions, in different contexts and even with different visual manifestations. Still, it is constantly used as a chronological or cultural marker in contemporary archaeology. The idea behind this article is to extract the available data concerning the WPP and try to embed it into the existing chronological framework. We are aware of the limitations of our approach and the unpopularity of centring an article on a small and isolated feature of the material culture. Nevertheless, we consider worth pursuing weather there is a pattern in the appearance and distribution of the WPP. Can this pattern be used in addressing the neolithisation process in the peninsula? Can this pattern be used for testing other elements of the material culture? An answer to these questions is the objective of this paper.

1. Time and territory

The early Neolithic ceramic assemblages from the Balkans exhibit high variety of techno–typological characteristics. What add even more to the variability are the combinations and the quantitative relations between these characteristics. Beside WPP, the assemblages include dark – painted motifs, different modes of impresso, plastic applications, but in general, the 'plain' or 'monochrome' pottery is dominant. WPP is usually a very small part of an assemblage – from one shard only (Stojanovski et al., 2014) to few percents. The paint is highly calcareous (marl) clay, applied on a vessel surface before firing (Rye, 1981; Libšer and Vilert, 1989; Чохаджиев, 2007; Dzhanfezova et al., 2014).

The most southern appearance of WPP is in the region of Central Macedonia, West of Thessaloniki (northern Greece).[1] Nea Nikomedeia and Giannitsa B are the first Neolithic settlements in the peninsula, which include WPP as part of their assemblage (group I in fig. 1). Today they are more than 30 km inland, but as geomorphology studies show, during the 7th and 6th millennium BC these locations were more or less on the coast of the Thermaic Gulf, which was stretching further to Northwest from its modern coast (Bintliff, 1976; Ghilardi et al., 2008). While in Nea Nikomedeia WPP is outnumbered by the dark-painted, in Giannitsa the numbers are reversed. Nevertheless, in both cases the amount of painted pottery is only few percent (Tsirtsoni, 2009). Absolute dates are available from Nea Nikomedeia (fig. 2). These dates have been elaborated by Reingruber and Thissen (2009), according to which the settlement was inhabited during the 63rd – 62nd century BC.

In Republic of Macedonia, two distinct early Neolithic groups are known.[2] The first one (group II in fig. 1) is spread in the eastern and northern part of the country and includes WPP sites from Ovče Pole (Barutnica – Anzabegovo and Nemanjica; Naumov et al., 2009; Санев, 2009), Skopje valley (Cerje – Govrlevo; Fidanoski, 2012) and Polog valley (Stenče and Dolno Palčište; Naumov et al., 2009). The other group (group V in fig. 1) includes settlements from the southern part of the country in the big, but relatively isolated Pelagonia valley (Veluška Tumba, Tumba Porodin, Tumba Optičari, Tumba Mogila, Čuka Topolčani and Vrbjanska Čuka; Naumov et al., 2009). Among other things (settlement pattern and duration, vessel shape), these two groups distinguish themselves by the ornamentation style of the WPP and the chronology. While group II lived in the last century of the 7th and the first two centuries of the 6th millennium[3] (fig. 3; Thissen, 2009; Fidanoski, 2012), the available dates

---

[1] The occurrence of WPP in Thessaly is extremely scarce and never in layers earlier than the WPP occurrence to the North (Tsirtsoni 2009).
[2] A third one (Zlastrana) is considered in the Lake Ohrid area, but is poorly known.
[3] Dates are available from Barutnica and Cerje.

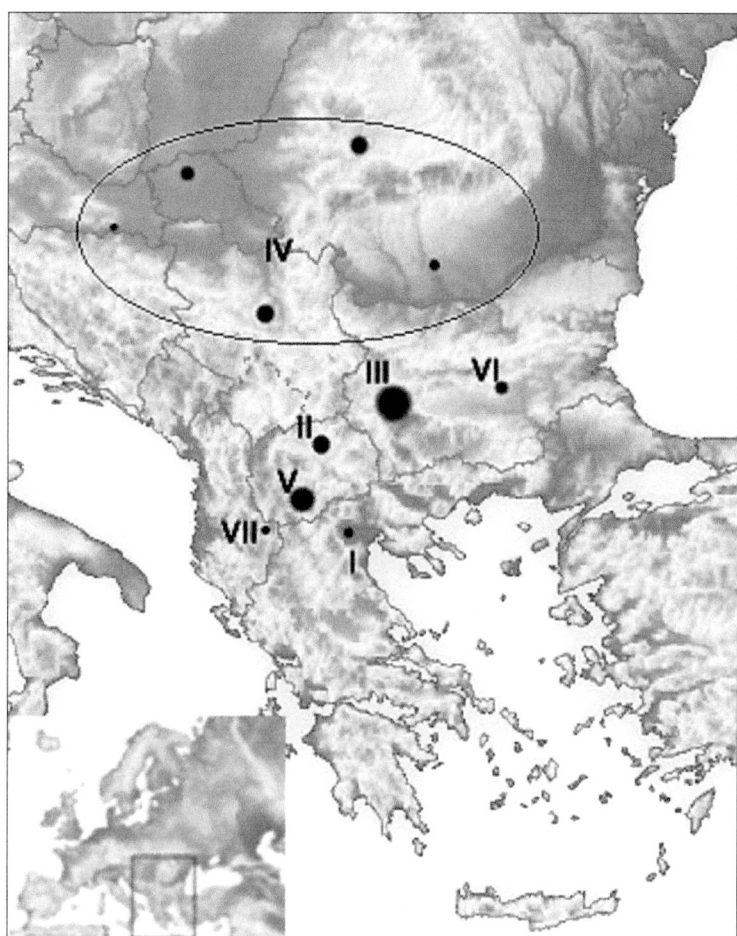

FIGURE 1. WHITE-PAINTED POTTERY (WPP) GROUPS IN THE BALKAN PENINSULA.

from four sites in group V (fig. 6) show that the tell settlements in Pelagonia were established not earlier than the 59th century BC, i.e. at the end of the early Neolithic phase of group II, and continued to produce the WPP even after the previous group developed middle Neolithic traits.

The middle Struma valley in Bulgaria has the highest concentration of sites with WPP, with a chronological span from the very beginning of the 6th millennium until the middle of it (group III in fig. 1; fig. 4). From the border with Greece to Pernik, several sites have been excavated and published: Kovačevo, Ilindenci, Elešnica, Vaksevo, Nevestino, Krainici, Galabnik, Negovanci, Priboy (Lichardus-Itten, 2009; Salanova, 2009; Grebska-Kulova et al., 2011; Чохаджиев et al., 2007; Чохаджиев, 1990; Чохаджиев, 2007). A bit further to the northeast, on the Sofia plateau are Slatina and Kremikovci (Nikolov, 2007; Тодорова and Вайсов, 1993). To this group we can add Džuljunica, the only site in Bulgaria north of Stara Planina where, according to the stratigraphy proposed by Elenski, WPP was found in layer II (Еленски, 2006; Krauß et al., 2014). As the plot in fig. 4 show, the developments in middle Struma and in Džuljunica II are more or less contemporary, while the Sofia area (as the Slatina dates show) was settled somewhat later.

Another known group of WPP sites from Bulgaria is in the Thracian valley (group VI in fig. 1). Besides the famous Karanovo tell, there are also the nearby Kazanluk and Azmak, as well as Kapitan Dimitrievo on the Southwestern edge of the valley. The available dates in fig. 7 are from Karanovo II, following in time Karanovo I (Тодорова and Вайсов, 1993). Nevertheless, because of the continuity between the two phases, it is safe to assume that the mound was established not earlier than the 58th millennium BC. The relations on stylistic ground with the Sofia plateau sites reconfirm this chronological position.

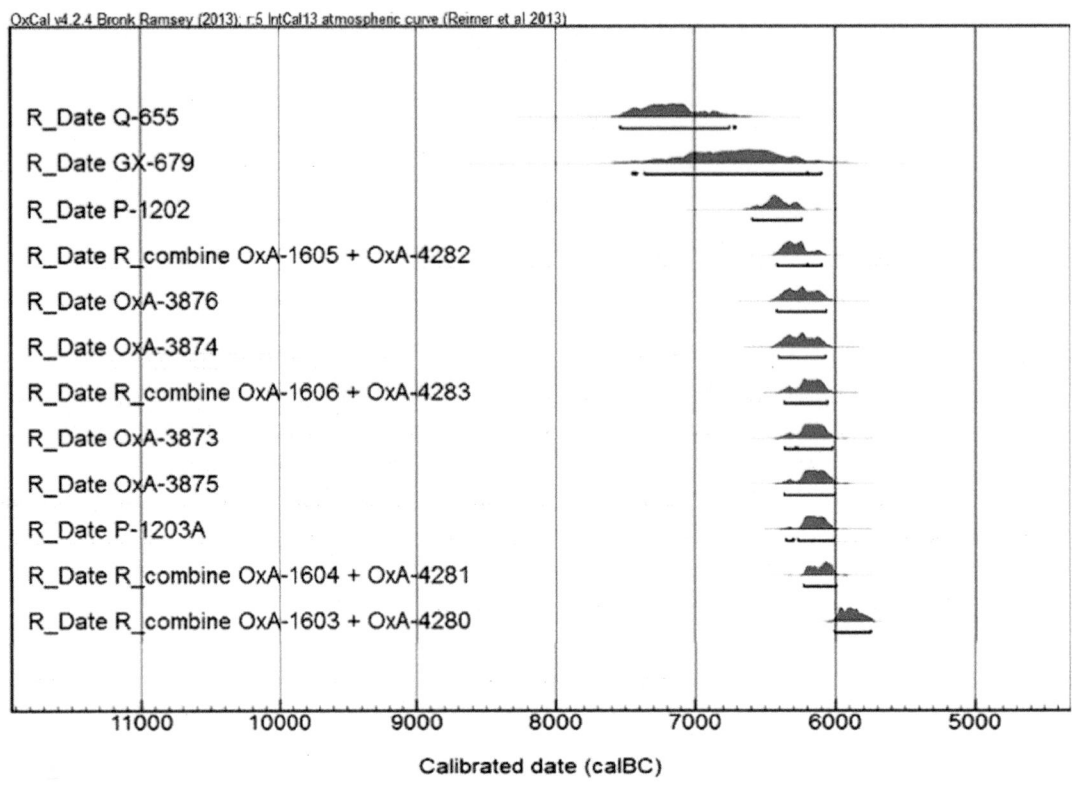

FIGURE 2. CALIBRATION PLOT FOR NEA NIKOMEDEIA (GROUP I); REFERENCES: (REINGRUBER AND THISSEN, 2009).

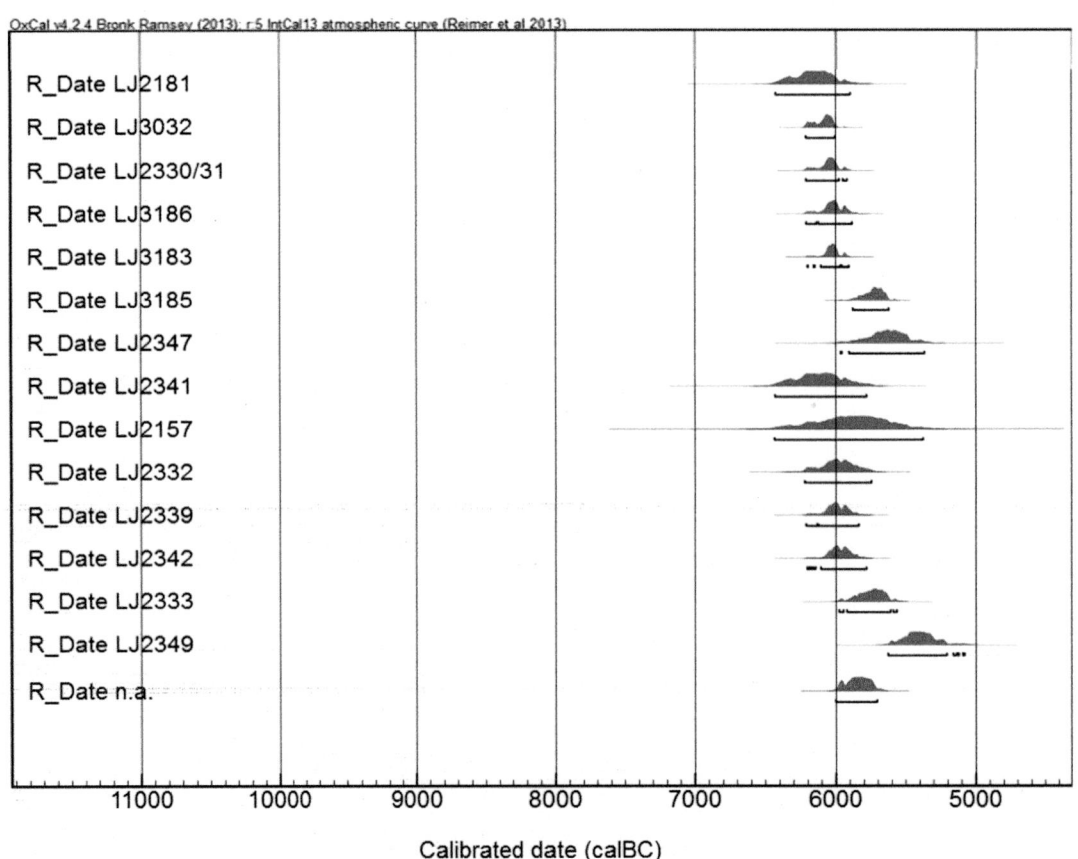

FIGURE 3. CALIBRATION PLOT FOR GROUP II (ANZABEGOVO IA: LJ 2181 – LJ 2347; ANZABEGOVO IB: LJ 2341 – 2349; GOVRLEVO: N.A. (LAB CODE UNAVAILABLE)); REFERENCES: (THISSEN, 2009; FIDANOSKI, 2012).

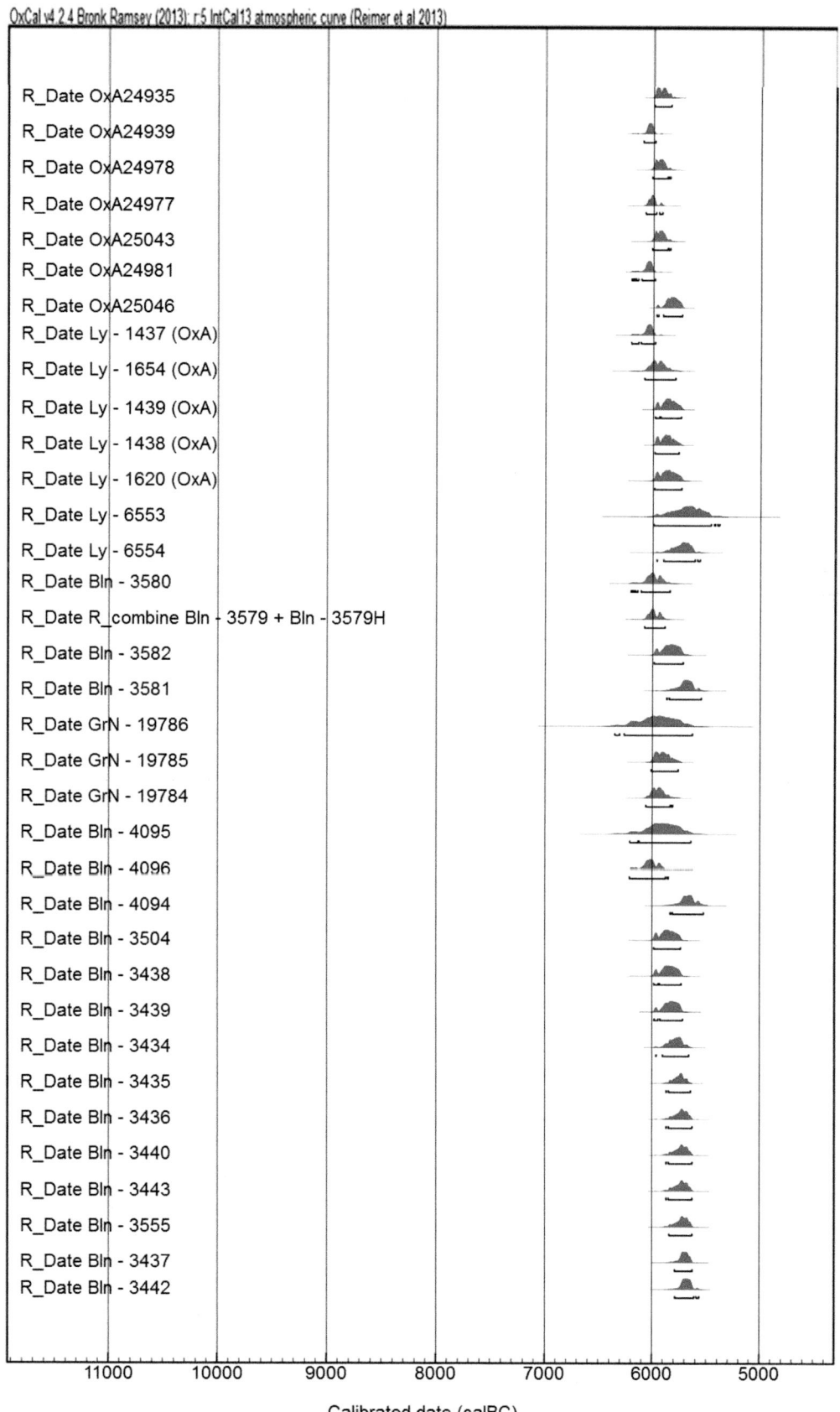

FIGURE 4. CALIBRATION PLOT FOR GROUP III: DŽULJUNICA II (OxA 24935 – OxA 25046); KOVAČEVO IA (LY – 1437 (OxA)); KOVAČEVO IA-IB (LY – 1654 (OxA)); KOVČEVO IB (LY – 1439 (OxA), LY – 1438 (OxA), LY – 1620 (OxA)); KOVAČEVO ID – II (LY – 6553, LY – 6554); GALABNIK (BLN-3580 – BLN-4094); SLATINA (BLN-3504 – BLN-3442); REFERENCES: (KRAUSS ET AL., 2014; ФИДАНОСКИ, 2013).

In the northern part of the Balkan Peninsula there are geographically dispersed sites with similar chronology (fig. 5), in which WPP was found. In fig. 1 they are united as group IV. Contemporary borders have separated them into three different modern countries. Few sites are grouped in Central Serbia: Ajman, Divostin, Grivac and Brdo (Тасић, 2009; Garašanin, 1979; Bogdanovich, 2007).

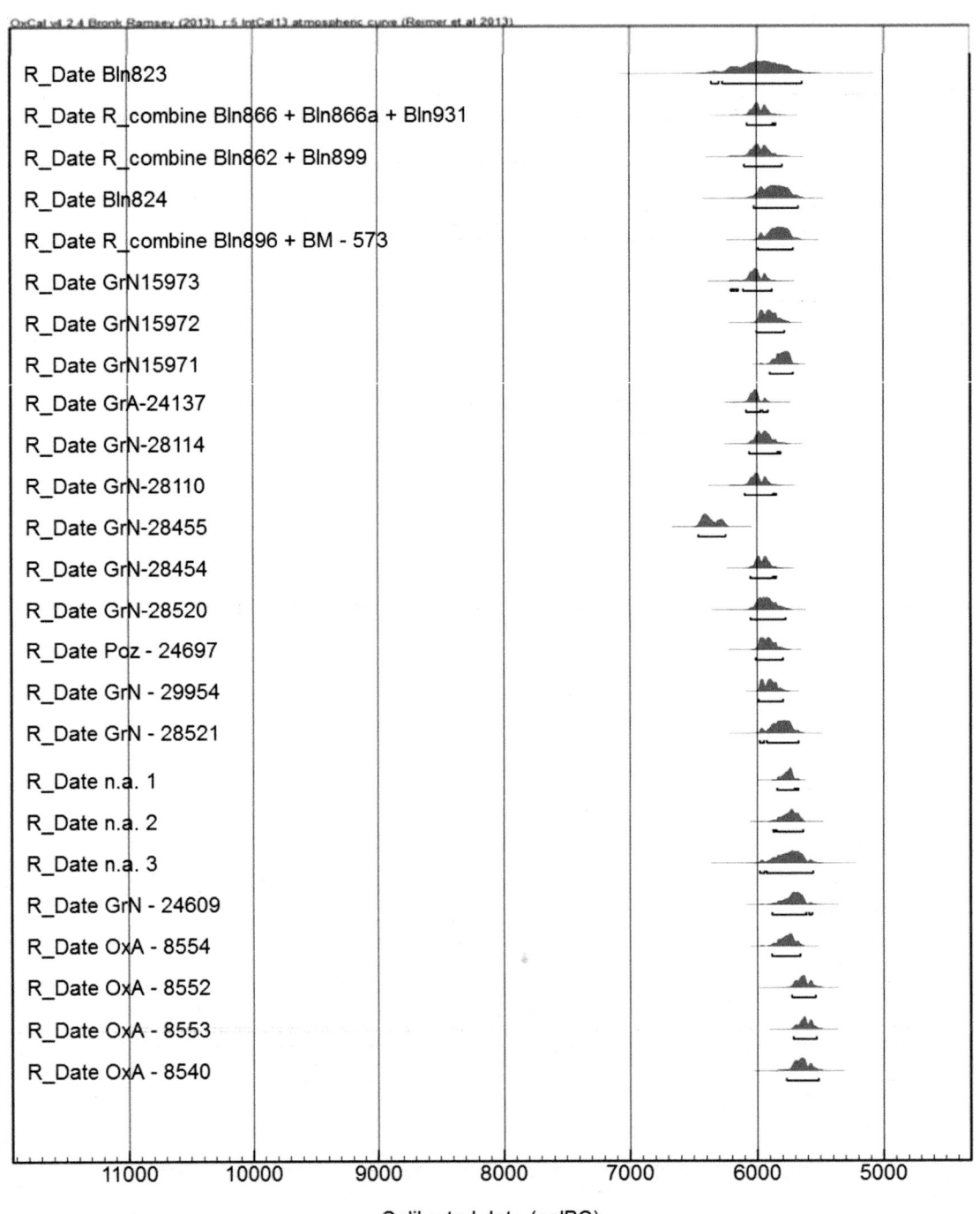

FIGURE 5. CALIBRATION PLOT FOR GROUP IV: DIVOSTIN II (BLN-823 – R_COMBINE BLN896 + BM-573); MAGAREČI MLIN (GRN 15973 – GRN 15971); GURA BACIULUI (GRA-24137); SEUSA (GRN-28114); OCNA SIBIULUI (GRN-28110); FOENI SALAS (GRN-28455, GRN-28454); MIERCUREA SIBIULUI – PETRIS (GRN-28520 – GRN-28521); GALOVO (N.A.1, N.A.2, N.A.3 (LAB CODES NOT UNAVAILABLE)); DONJA BRANJEVINA II (GRN-24609); BISERNA OBALA – NOSA (OXA-8554 – OXA-8553); LUDOŠ – BUDŽAK (OXA-8540); REFERENCES: (THISSEN, 2009; KARMANSKI, 2005; LUCA ET AL., 2011; MINICHREITER, 2007; BIAGI ET AL., 2005; ТАСИЋ 2009).

Dates are available and presented in fig. 5 only from Divostin II. Four sites are known in Vojvodina: Donja Branjevina (layer II), Magareči Mlin, Ludoš – Budžak and Sajan (Karmanski, 2005; Тасић, 2009). The first three are represented with ¹⁴C dates in fig. 5. To the West from here, in Croatia, is the site Galovo (fig. 5; Minichreiter, 2007). Chronologically, geographically and stylistically, this assemblage is very close to Donja Branjevina II. East of Vojvodina to Transylvania, several sites have been discovered and dated: Foeni – Sălaş, Miercurea Sibiului, Ocna Sibiului, Şeuşa and Gura Baciului (Luca *et al.*, 2011). In South Romania, the area between the South Carpathians and the Danube, three WPP sites were discovered: Grădinile, Cârcea and Măgura (Andreescu and Mirea, 2008; Luca *et al.*, 2011). Dates are lacking so far from these sites, but as far as material culture is considered, they can be associated with certainty with the Transylvanian sites, as well as with Džuljunica II across the Danube and the sites from the Struma valley in Bulgaria.

Two WPP sites exist in the extreme East of Albania, near the border with Macedonia and Greece: Podgorie and Vashtëmi (Korkuti, 2007; Bunguri, 2014). Here they are gathered in group VII (fig. 1). Unfortunately dates are not available. Considering the geographical position and the complexity of the motifs, closest associations can be pursued in the assemblages from the sites in Pelagonia.

2. Discussion

So far we have presented a network of early Neolithic sites in the Balkans, where WPP has been discovered. Within this network, seven territorially integrated, but still arbitrary groups can be distinguished (fig. 1). We do not claim that all WPP sites that exist in the peninsula are presented in the groups. There is no strict chronological, neither cultural dimension in their separation. The sites have their own life duration and dynamics, and not all of them continue their post-WPP life into the local middle or late Neolithic cultures. The unifying feature is the fact that at some sequence, or during their entire existence, WPP was part of the assemblage. With the available ¹⁴C dates we have today from the entire time span of existence of the WPP (fig. 2 – 7), we can distinguish three stages.

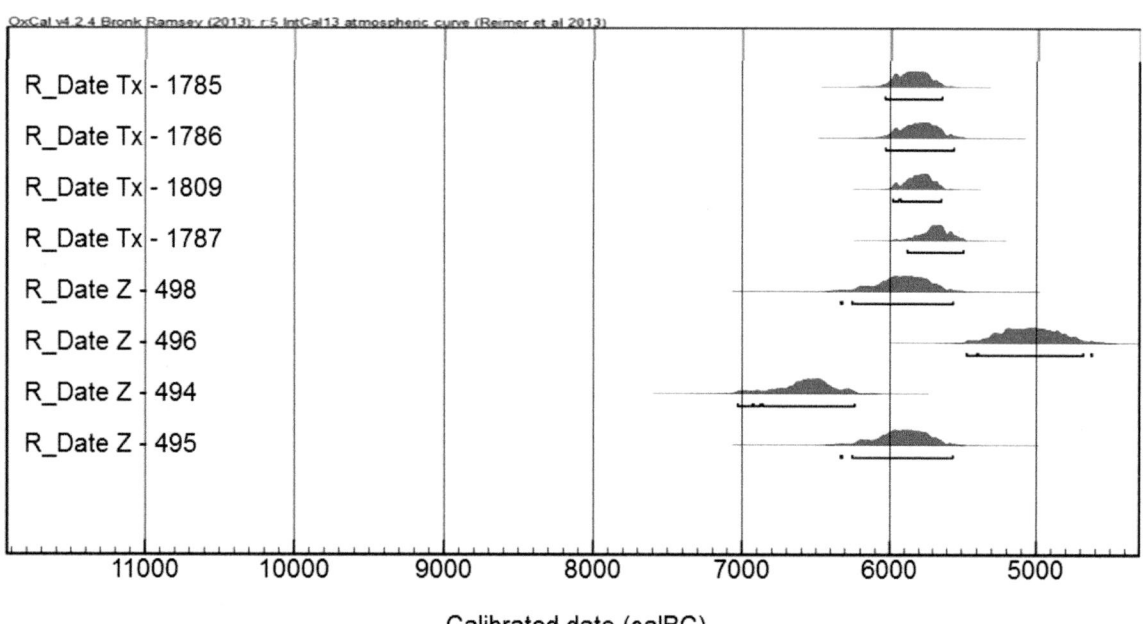

FIGURE 6. CALIBRATION PLOT FOR GROUP V: VELUŠKA TUMBA (TX-1785 – TX-1809);
TUMBA PORODIN (TX-1787); TUMBA MOGILA (Z-498, Z-496); ČUKA TOPOLČANI (Z-494, Z-495);
REFERENCES: (SRDOČ *ET AL.*, 1977; NAUMOV *ET AL.*, 2009).

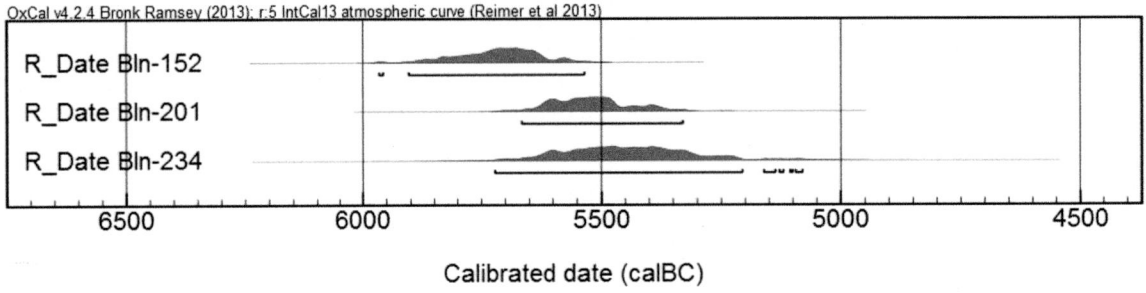

FIGURE 7. CALIBRATION PLOT FOR KARANOVO II (GROUP VI); REFERENCES: (ТОДОРОВА AND ВАЙСОВ, 1993).

The first one is represented by group I – the sites in Northern Greece, where this fashion in pottery appears for the first time in the peninsula. The obvious question is how this decoration appeared; if we consider exogenous source, then from where and in what shape (an object or an idea) was it imported? By the time when Nea Nikomedeia was settled, the assemblages in the Anatolian Lake District area already contained significant percentage of pottery with painted decoration (Krauß, 2011, p. 110). The influence of Hacilar over the neolithisation process of Southeast Europe has been noted by Brami and Heyd (2011, p. 178). Agathe Reingruber (2011) also points to links between the two regions, more specifically in the cultivated plant species and the use of clay stamps (Reingruber, 2011, pp. 300-301). In her view of the Aegean during the transformation process from Mesolithic to Neolithic – a very dynamic world with seafaring people moving back and forth in various directions across the sea, different ideas would find different fertile ground at different time. Somehow in this complex and 'regionalized' pattern (or rather lack of pattern) of neolithisation, the WPP style appeared in the north-western fringes of the sea.

Almost two hundred years later, during the last century of the 7th millennium, this decoration style appeared 140 km to the North in Ovče Pole (group II). This marks the beginning of a second stage, during which a more rapid and multidirectional dispersal occurred. Within a century after this event, similar decoration patterns appeared in the Struma valley and Džuljunica (group III), and Central Serbia and Romania (group IV). The concept of pottery decoration is the same, but there is great diversity in motifs and contexts.

The period after 5800 calBC marks the third stage. Several developments are evident during this stage. The previously established sites in group II introduced new, more elaborated vessel shapes and dark paint for the ornaments – traits associated with the middle Neolithic phase in Macedonia. The WPP slowly disappears (Гарашанин and Гарашанин, 2009). Similar picture emerges from some sites in the Struma basin and Džuljunica (Чохаджиев, 2007; Еленски, 2006), even though the chronological system used here is different (Тодорова and Вайсов, 1993). For our subject here, more significant is the continuation of the spread of WPP in new areas. At this point, or just slightly earlier, the settlements in the Sofia plateau and in Thracian valley appeared (group VI). More or less contemporary is the establishment of the tell settlements in Pelagonia (group V), the later occurrences in group IV (the sites in Vojvodina and Galovo), and probably this is when the two sites in Albania were settled (group VII).

3. Conclusion

From the emerging pattern, several things can be suggested in the context of the process of neolithisation of Southeast Europe. After a period of 'fermentation' on the Aegean coast, WPP appeared in newly settled sites in the northern hinterland. Obviously, natural communication lines – like Vardar and Struma valleys – played a role in this initial dispersal, which was relatively rapid, and affected areas all the way to the Carpathian Mountains. The variety of archaeological contexts in which WPP appears, as well as the variations in the motifs of the decoration, have led to several systematisations in the past and are the base for the establishment of the many regional early Neolithic cultures (Garašanin, 1979; Чохаджиев, 1990; Чохаджиев, 2007; Nikolov, 2007; Тасић, 2009). It is difficult to imagine groups of immigrants coming from outside, with a complete set of material culture and immediately establishing successful communities and communication networks in previously uninhabited space. The presented so far, in our view is more in tone with a change in the life-style of communities already present in the territory, familiar with the environment and resources (Thissen, 2005). When these communities were introduced to Neolithic novelties (among which was the WPP), they adopted them at different pace and manner, in accordance with their own necessities, traditions and worldview. Probably these novelties arrived with the seafaring foreigners, who must have been familiar through previous contacts. Maybe some of them eventually decided to stay and form small communities, but the main driving force behind the spread of the WPP and the Neolithic in general, would have been the local communities. At this critical point in time, what stood behind the formation of the Neolithic cultures was probably a complex system of interactions between societies with different provenance, mobility level, subsistence patterns and archaeological visibility. The stage revealed before us through excavation, might actually reflect the result of a unifying tendency during and after the initial impulse of cultural interchange. We do not know the exact mechanisms of cultural homogenizations, but, considering the speed with which the unifying cultural traits were spread, polarization or conflict can be excluded.

There is much work to be done in the Balkan Peninsula, before we can reveal a more detailed pattern of appearance, development and transformation of the Neolithic communities and the social network of communication between them. The one emerging from the tracking of the WPP, with the limitations noted in the introduction, is still very general. Nevertheless, it provokes new questions, which in our view are worth following.

Acknowledgments

This article suffered many changes since the first version, trying to make more sense from the matter at hand. If it succeeded, the constructive criticism of Laurens Thissen helped a lot. If it did not, the fault is of the author. An early version was presentation at the 19th UISPP conference in Burgos. The author's participation in the conference was financially supported by an Erasmus Mundus PhD grant, in the frames of the IDQP program. Gratitude is also due to Stefan Chohadziev and Veneta Genadieva for making possible the observation of early Neolithic collections in the archaeological museum in Kjustendil.

References

Andreescu, R.-R. and Mirea, P., 2008 – Teleorman Valley. The beginning of the Neolithic in Southern Romania. Acta Terrae Septemcastrensis, VII.
Biagi, P., Shennan, S. and Spataro, M., 2005 – Rapid rivers and slow seas? – New data for the radiocarbon chronology of the Balkan Peninsula. In: Nikolova, L.; Fritz, J. and Higgins, J., eds. Prehistoric Archaeology and Anthropological Theory and Education. pp. 41-50.
Bintliff, J., 1976 – The Plain of Western Macedonia and the Neolithic Site of Nea Nikomedeia. Proceedings of the Prehistoric Society, 42, pp. 241-262.
Bogdanovich, M., 2007 – Proto-Starchevo Culture and Early Neolithic in the Struma Valley. In: Stefanovich, M.; Ivanov, G. and Todorova, H., eds. Strymon Praehistoricus.
Brami, M. and Heyd, V., 2011 – The origins of Europe's first farmers: The role of Hacılar and western Anatolia, fifty years on. Praehistorische Zeitschrift, 86(2), pp. 165-206.
Bunguri, A., 2014 – Different models for the Neolithisation of Albania. Documanta Praehistorica, 41, pp. 79-94.
Çilingiroglu, Ç., 2005 – The concept of 'Neolithic package': considering its meaning and applicability. Documenta Praehistorica, 32, pp. 1-13.

Dzhanfezova, T., Doherty, C. and Elenski, N., 2014 – Shaping a future of painting : the early Neolithic pottery from Dzhulyunitsa, North Central Bulgaria. Bulgarian e-Journal of Archaeology, 4, pp. 137-159.

Fidanoski, L., 2012 – Cerje – Govrlevo and Miloš Bilbija, Skopje: Museum of the city of Skopje.

Garašanin, M., 1979 – Centralnobalkanska zona. In: Benac, A., ed. Praistorija Jugoslavenskih Zemalja II, Neolitsko doba. Sarajevo: Akademija nauke i umetnosti Bosne i Hercegovine, pp. 79-212.

Ghilardi, M. *et al.*, 2008 – Human occupation and geomorphological evolution of the Thessaloniki Plain (Greece) since mid Holocene. Journal of Archaeological Science, 35, pp. 111-125.

Grebska-Kulova, M. *et al.*, 2011 – Ранонеолитното селище при с. Илинденци, м. Масовец, община Струмяни. Археология.

Karmanski, S., 2005 – Donja Branjevina: A Neolithic settlement near Deronje in the Vojvodina (Serbia) quaderno – Biagi, P., ed., Trieste: Società per la Prehistoria e Protostoria della regione Friuli-Venezia Giulia.

Korkuti, M., 2007 – The Early Neolithic of Albania in a Balkan Perspective. In: Spataro, M. and Biagi, P., eds. A short walk through the Balkans: The first farmers of the Carpathian basin and adjacent regions. Trieste: Società per la Prehistoria e Protostoria della regione Friuli-Venezia Giulia, pp. 113-119.

Krauß, R. *et al.*, 2014 – Beginnings of the Neolithic in Southeast Europe: the Early Neolithic sequence and absolute dates from Džuljunica-Smardeš (Bulgaria). Documenta Praehistorica, 41, pp. 51-78.

Krauß, R., 2011 – On the 'Monochrome' Neolithic in Southeast Europe. In: R. Krauß, ed. Beginnings – New Research in the Appearance of the Neolithic between Northwest Anatolia and the Carpathian Basin Papers of the International Workshop 8th – 9th April 2009, Istanbul Organized by Dan Ciobotaru, Barbara Horejs and Raiko Krauß. pp. 109-125.

Libšer, I. and Vilert, F., 1989 – Tehnologija keramike III., Belgrade: Univerzitet Umetnosti u Beogradu.

Lichardus-Itten, M., 2009 – La néolithisation des Balkans méridionaux vue à travers la céramique de Kovačevo. In: Astruc L., Alain, G. and Salanova, L., eds. Méthodes d'approche des premières productions céramiques: étude de cas dans les Balkans et au Levant Table-ronde de la Maison de l'Archéologie et de l'Ethnologie (Nanterre, France) 28 février 2006.

Luca, S. A., Suciu, C. I. and Dumitrescu-Chioar, F., 2011 – Starčevo – Criş Culture in Western part of Romania – Transylvania, Banat, Crişana, Maramureş, Oltenia and Western Muntenia: repository, distribution map, state of research and chronology. In: Luca, S. A. and Suciu, C. I., eds. The first Neolithic sites in Central/South-East European transect. Vol. 2: Early Neolithic sites on the territory of Romania. Krakow: Archaeopress.

Minichreiter, K., 2007 – The White-painted Linear A Phase of the Starčevo Culture in Croatia. Pril. Inst. Arheol. Zagrebu, 24, pp. 21-34.

Naumov, G. *et al.*, 2009 – Neolithic communities in the Republic of Macedonia, Skopje: Dante.

Nikolov, V., 2007 – Problems of the early stages of the neolithisation in the Southeast Balkans. In: Spataro, M. and Biagi, P., eds. A Short Walk through the Balkans: the First Farmers of the Carpathian Basin and Adjacent Regions. Trieste: Società per la Prehistoria e Protostoria della regione Friuli-Venezia Giulia, pp. 183-188.

Özdoğan, M., 2011 – Archaeological Evidence on the Westward Expansion of Farming Communities from Eastern Anatolia to the Aegean and the Balkans. Current Anthropology, 52(S4), pp. S415-S430. Available at: http://www.jstor.org/stable/info/10.1086/658895 [Accessed March 20, 2014].

Özdoğan, M., 2010 – Westward expansion of the Neolithic way of life: sorting the Neolithic package into distinct packages. In: Matthiae, P. *et al.*, eds. Near Eastern archaeology in the past, present and future: heritage and identity, vol. 1 of Proceedings of the 6th International Congress of the Archaeology of the Ancient Near East. Wiesbaden, Germany: Harrassowitz., pp. 883-897.

Perlès, C., 2001a – An alternate (and old-fashioned) view of Neolithisation in Greece. Documenta Praehistorica, 30(495), pp. 99-113.

Perlès, C., 2001b – The Early Neolithic in Greece. The First Farming Communities in Europe, Cambridge University Press.

Reingruber, A., 2011 – Early Neolithic settlement patterns and exchange networks in the Aegean. Documenta Praehistorica, XXXVIII, pp. 291-305.

Reingruber, A. and Thissen, L., 2009 – Depending on ^{14}C data: chronological frameworks in the Neolithic and the Chalcolithic of Southeastern Europe. Radiocarbon, 51(2), pp. 751-770.

Rye, O. S., 1981 – Pottery technology: principles and reconstruction, Taraxacum.

Salanova, L., 2009 – La plus ancienne céramique bulgare (Kovačevo, Bulgarie): caractérisation technique, implications socio-culturelles. In: Astruc, L.; Gaulon, A. and Salanova, L., eds. Méthodes d'approche des premières productions céramiques: étude de cas dans les Balkans et au Levant. Table-ronde de la Maison de l'Archéologie et de l'Ethnologie (Nanterre, France) 28 février 2006. Rahden/Westf.: Verlag Marie Leidorf GmbH, pp. 21-28.

Srdoč, D. et al., 1977 – Rudjer Boškovič Institute radiocarbon measurements IV. Radiocarbon, 19.

Stojanovski, D., Nacev, T. and Arzarello, M., 2014 – Pottery typology and the monochrome Neolithic phase in the Republic of Macedonia. In: Schier, W. and Drasovean, F., eds. The Neolithic and Eneolithic in Southeast Europe – New Approaches to Dating and Cultural Dynamics in the 6th to 4th Millenium BC. Rahden/Westf.: Verlag Marie Leidorf GmbH, pp. 9-27.

Thissen, L., 2009 – First Ceramic Assemblages in the Danube catchment, SE Europe – a Synthesis of the Radiocarbon Evidence. Teleorman County Museum Bulletin, (1), pp. 9-30.

Thissen, L., 2005 – The role of pottery in agropastoralist communities in early Neolithic southern Romania. In: Bailey, D.; Whittle, A. and Cummings, V., eds. (un)settling the Neolithic. Oxbow Books, pp. 71-78.

Tsirtsoni, Z., 2009 – A question of status: interpreting ceramic variability in Early Neolithic Northern Greece. In: Astruc, L.; Gaulon, A. and Salanova, L., eds. Méthodes d'approche des premières productions céramiques: étude de cas dans les Balkans et au Levant. Table-ronde de la Maison de l'Archéologie et de l'Ethnologie (Nanterre, France) 28 février 2006. Rahden/Westf.: Verlag Marie Leidorf GmbH, pp. 39-50.

Гарашанин, М. and Гарашанин, Д., 2009 – Керамика. In В. Санев, ed. Анзабегово-населба од Раниот и Среден Неолит во Македонија. Штип: Завод за заштита на споменицте на културата и музеј – Штип.

Еленски, Н., 2006 – Сондажни проучвания на ранонеолитното селище Джулюница-Смърдеш, Великотърновско (предварително съобщение). Археология, 47, pp. 96-117.

Санев, В., 2009 – Анзабегово-населба од Раниот и Среден Неолит во Македонија, Штип.

Тасић, Н. Н., 2009 – Неолитска квадратура круга, Београд: Завод за уџбенике.

Тодорова, Х. and Вайсов, И., 1993 – Новокаменната епоха в България, София.

Фиданоски, Љ., 2013 – Раниот Неолит во Македонија и Бугарија: географски контакти и културни релации. Славишки зборник, 2.

Чохаджиев, М., 1990 – Ранния неолит в Западна България – поява, развитие и контакти. Proceedings of the museum of history Kyustendil, 2.

Чохаджиев, С., 2007 – Неолитни и халколитни култури в басейна на река Струма, Велико Трново.

Чохаджиев, С., Бакъмска, А. and Нинов, Л., 2007 – Крайници – ранокерамичното селище от басейна на река Струма. In: Stefanovich, M.; Ivanov, G. and Todorova, H, eds. Strymon Praehistoricus. Gerda Henkel Stiftung.

Settlements – Head and Settlements – Tail in the Neolithic Obsidian Exchange Network in the Western Mediterranean

Tania Quero
Specialisation School in Archaeological Heritage –
University of Venezia, Udine, Trieste

Abstract
The obsidian exchange is one of the most important examples of long-distance trade in Europe, during the Neolithic period, because of its social and economic role. This paper will evaluate two Neolithic sites in Northern and Southern Italy, in terms of different participation to the obsidian exchange network.
Key words: *Italy, Western Mediterranean, Neolithic, Obsidian, Exchange*

Résumé
L'échange de l'obsidienne est un exemple important de circulation sur longue distance en Europe, pendant le Néolithique, grâce à son rôle social et économique. On y évaluera deux sites au nord et au sud de l'Italie, au regard de leur participation différente au réseau d'échange de l'obsidienne.
Mots clés: *Italie, Méditerranée occidentale, Néolithique, Obsidienne, échange*

Introduction

The study of the obsidian circulation in the Western Mediterranean sea, during the Neolithic period, is very interesting because of its important social and economic role in long-distance trading (Ammerman, 1985, pp. 81-82; Tykot, 2001, p. 25).

This contribution would analyse the different participation to the obsidian exchange networks of two sites in Italy, considering their distance from the sources and the competition with other raw materials.

The sites

At S. Martino Spadafora site (fig. 1), on the Northern coast of Sicily (Southern Italy), two occupation phases are attested during the Early Neolithic period (Stentinello culture: 4850-4540 cal BC, 4720-4480 cal BC) and Late Neolithic period (Diana culture: 4240-3970 cal BC) (Martinelli, Quero, 2013, p. 813).

In both the phases, the obsidian comes from Lipari Island (XRF analysis by University of Ferrara) and covers the 98% of the whole lithic assemblage. There are almost all stages of the reduction sequence (fig. 2).

However, it is possible to identify a diachronic change in the obsidian exploitation.

- In the ancient occupation, the cores are overexploited, the flakes production exceeds the laminar blanks (probably obtained by indirect percussion).
- In the second occupation, the cores have standardized shapes, flakes and blades (obtained by pressure technique) have about the same percentage.

Flakes are preferred blanks for tools. Laminar blanks are often unretouched (fig. 3).

The Granarolo site dates at the transition from the first to the second phase of the Square Mouth Pottery culture (5700-5300 BP – Middle Neolithic period). It lies (fig. 1) in the central Po valley,

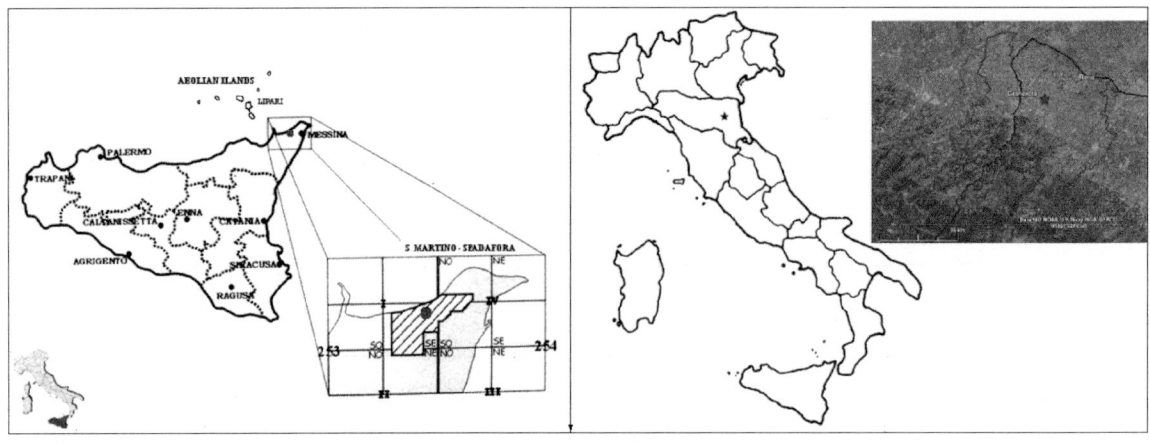

FIGURE 1. SITES POSITION: S. MARTINO SPADAFORA ON THE LEFT,
GRANAROLO DELL'EMILIA ON THE RIGHT.

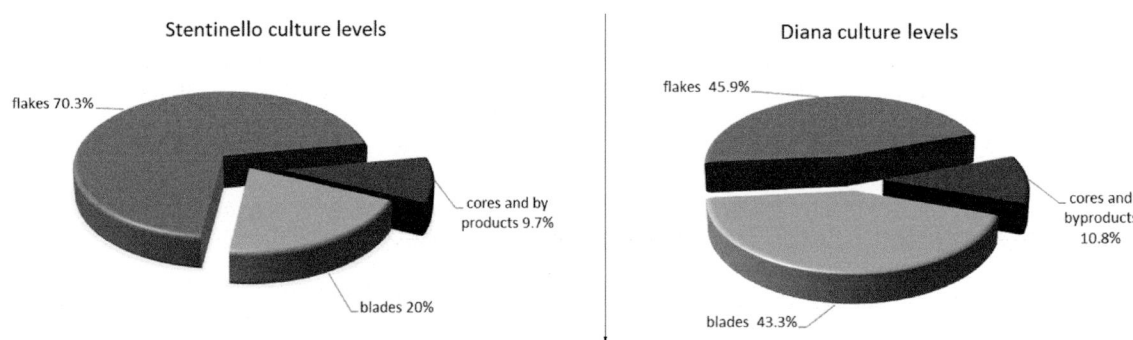

FIGURE 2. S. MARTINO SPADAFORA SITE: TECHNOLOGICAL DISTRIBUTION OF
THE OBSIDIAN KNAPPING PRODUCTS IN TWO OCCUPATION PHASES.

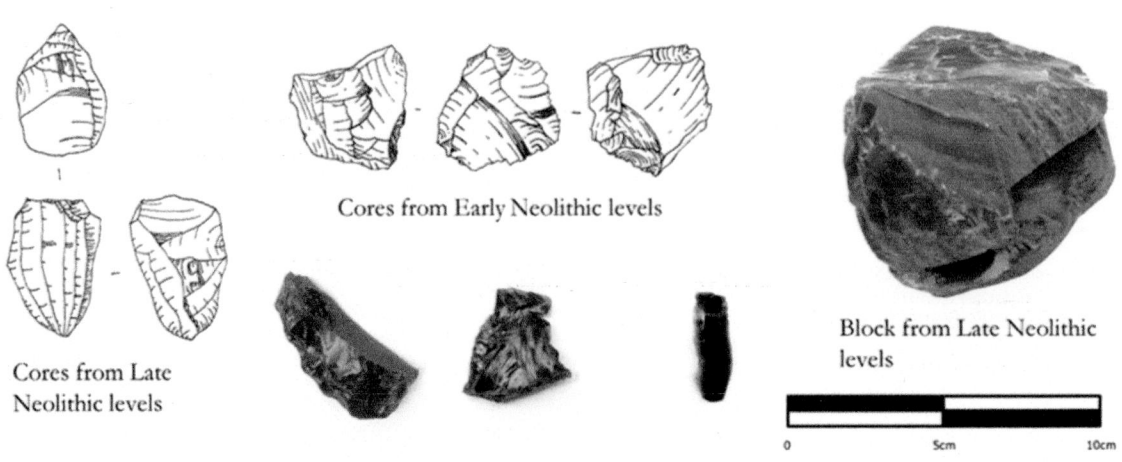

FIGURE 3. S. MARTINO SPADAFORA SITE: OBSIDIAN CORES AND PRODUCTS.
TOOLS ARE SCRAPERS, DENTICULATES AND PERFORATORS.

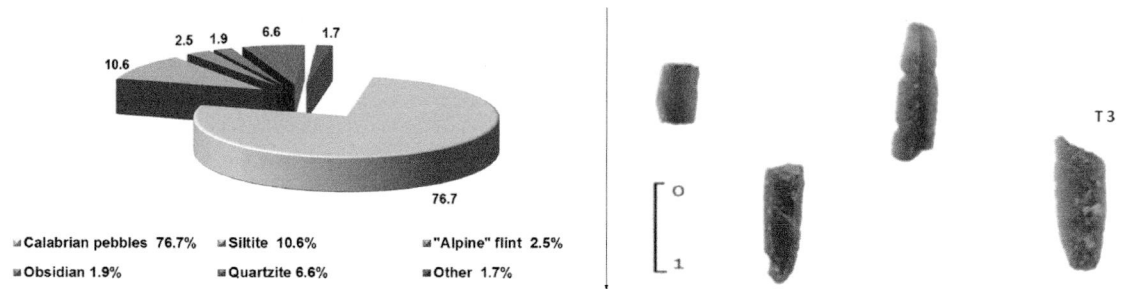

FIGURE 4. GRANAROLO SITE: RAW MATERIALS DISTRIBUTION AND THE KNAPPING PRODUCTS.
THE BLADELETS ARE UNRETOUCHED: THERE ARE JUST A TRANSVERSAL BURIN AND A TRUNCATURE.

near Bologna (Northern Italy). The flint provisioning is mostly local. Among exotic materials the obsidian, which the source is still unknown, accounts for the 1.9% of the whole assemblage (fig. 4): there are only 10 bladelets obtained by pressure technique, byproducts and cores are absent (Quero, 2013, p. 52).

Results

In the S. Martino site the obsidian exploitation changes from the Early to the Late Neolithic period. A few factors could be involved in this different behaviour: cultural choices, diversification in subsistence activities or intensive access to source and procurement.

Instead, the investigation at the Granarolo site confirms that also in the district of Bologna the Square Mouth Pottery culture communities had many contacts with the Central and Southern Italy, so that they could easily obtain some exotic materials, including the obsidian (Ammerman, Polglase, 2001, pp. 292-293; Tykot *et al.*, 2005, p. 105).

Conclusions

The obsidian exchange is still one of the most important examples of the long-distance trade in Europe, during the Neolithic period. Evidently, a raw material gains prestige and symbolic value when the distance from the source increases. The case of Granarolo is significant because non-local raw materials (as obsidian and 'alpine' flint) arrived in a place where local flint was preferred.

Otherwise, S. Martino Spadafora site possibly was part of the Lipari obsidian networks of exchange and had a role of distribution, thanks to its strategic location on the Northern coast of Sicily, being a bridge to the Calabria (Robb, 2003, pp. 956-957) and the Sicilian hinterlands (Ammerman, 1985, p. 98).

References

Ammerman, A. J., 1985 – The Acconia survey: Neolithic settlement and the obsidian trade. London, University of London, Institute of Archaeology. 10.

Ammerman, A. J.; Polglase, C., 2001 – Obsidian at Neolithic sites in Northern Italy. Preistoria Alpina. Trento. 34: pp. 291-296.

Martinelli, M. C.; Quero, T., 2013 – Cultural and Trade Networks in the Western Mediterranean, during the Neolithic Period: a Testimony from Northern Sicily. The S. Martino-Spadafora (ME) Site. Proceedings of the 16th Symposium on Mediterranean Archaeology. Florence, Italy, 1-3 March 2012. Oxford: B. A. R., pp. 813-822. (BAR International Series; 2581 II).

Quero, T., 2013 – L'industria litica di Granarolo dell'Emilia (BO) – via Foggia Nuova, nel quadro del pieno Neolitico dell'Italia settentrionale (cultura dei V.B.Q.). Unpublished MA thesis, University of Ferrara (Supervisors: Dr Federica Fontana, Dr Giuliana Steffè).

Robb, J., 2003 – Scavi neolitici e dell'età del Bronzo ad Umbro, Bova Marina, Reggio Calabria. Atti XXXV Riunione scientifica Istituto Italiano di Preistoria e Protostoria, Castello di Lipari 2-7 giugno 2000. Firenze. 35: pp. 955-958.

Tykot, R. H., 2001 – Neolithic exploitation and trade of obsidian in the central Mediterranean: new results and implications for cultural interaction. Acts of the XIVth UISPP Congress. University of Liège, Belgium, 2-8 September 2001. Oxford: B. A. R., pp. 25-35. (BAR International Series; 1303).

Tykot, R. H.; Ammerman, A. J.; Bernabò Brea, M. A.; Glascock, M. D.; Speakman R. J., 2005 – Source analysis and the socioeconomic role of obsidian trade in Northern Italy: New data from the Middle Neolithic site of Gaione. Geoarchaeological and Bioarchaeological Studies. 3: pp. 103-106.

Original and Skeuomorph: On the materiality of the Chalcolithic package of prestige in South Eastern Europe

Dragoş Gheorghiu
Doctoral School, National University of Art, Bucharest

Abstract
The establishment in South Eastern Europe of trade routes during the Neolithic and Chalcolithic allowed the circulation of a series of exotic materials from the south to the north of that region, which acquired a special value (i.e. social status) due to their rarity. As a consequence of their scarcity the exotic objects were copied in local materials, creating skeuomorphs. It seems that the symbolism of the materiality had a strong social impact, being associated with 'prestige'. The paper discusses the standardised prestige package of local and exotica materials in the Gumelniţa-Karanovo Kodjadermen and Cucuteni-Tripolye Chalcolithic traditions.

Key words: *Chalcolithic, Gumelniţa-Karanovo-Kodjadermen, Cucuteni-Tripolye, exotica, prestige package*

Résumé
La formation des routes de commerce dans le sud-est de l'Europe pendant le Néolithique et le Chalcolithique a permis la circulation du sud au nord d'une série de matériaux exotiques qui ont acquis une valeur spéciale (statut social) due à leur rareté. En conséquence, les objets exotiques ont été copiés dans les matériaux locaux, en créant des skeuomorphes. Il parait que le symbolisme de leur matérialité a eu un grand impact social, en étant associé au ' prestige '. Le chapitre présente les assemblages de prestige fait de matériaux locaux et exotiques dans les traditions chalcolithiques de Gumelniţa-Karanovo-Kodjadermen et Cucuteni-Tripolye.

Mots clés: *Chalcolithique, Gumelniţa-Karanovo-Kodjadermen, Cucuteni-Tripolye, exotica, paquet de prestige*

Introduction: Roads and materials in the Chalcolithic Gumelniţa-Karanovo-Kodjadermen tradition

The Chalcolithic in south-eastern Europe was an epoch when the Gumelniţa-Karanovo-Kodjadermen (GKK) complex expanded into the eastern Balkans, up to the area north of the Danube River (Popovici, 2010: 92). The Gumelniţa tradition spread northwards up to the hills of the Southern Carpathians (Frînculeasa, 2011), to the north of the Danube Delta (Bicbaev, 2010: 222), and south of Moldavia (Dragomir, 1983).

During this period the materiality of prestige items played an important role in the creation of social and exchange networks generated by a new stratified society (Renfrew, 1986; Chapman, 2013). One can observe a series of differences between the variety of the inventories of the southern large tell-settlements (Todorova, 1982) and that of the villages from the hilly sub-Carpathian region (Frînculeasa, 2011), which is indicative of the existence of an inter-settlements hierarchy.

In this respect the materialities from the Varna cemetery (Ivanov, 1988; Ivanov and Avramova, 2000) for example, could be an index of social inequality (see Renfrew, 1986) and the existence of trade routes for local (Rusev *et al.*, 2010), or exotic (Séfériades, 2010; Higham *et al.*, 2007) products within a politics of value of the emergent elites.

In GKK the strategy of social differentiation used a new dominant materiality which was that of metals (see Pernicka and Anthony, 2010), along the exotic valuable materials, such as shells, marble or chalcedony. A new materiality also meant a new technology, therefore the prestige and value of copper or gold objects can also be interpreted as a result of a new technological know-how.

The rise of the metal created new cultural models that started to circulate over relatively long distances between centre and peripheries, and Varna's position in the GKK could have created a 'Varna effect'

(Chapman, 2013) by disseminating its cultural models. Such a relationship between a centre and its margins (Guilaine, 2004) should be perceived as representing a two-stage process. In the first stage the centre acts as a (cultural) attractor by acquiring different materials, local or exotic, from various distances. At Varna the metal was procured from relatively close areas from the Balkans (Rusev *et al.*, 2010: 30), as opposed to the exotic materials coming from farther away (Séfériades, 2010).

After the technological processing of the valuable materials, some of the resulting objects and ideas were spread outside the tradition, to the northern area of Cucuteni-Tripolye. To approach this cultural process the present paper will discuss two issues identified in the process of the emergent Chalcolithic GKK stratified society: first, the existence of a prestige 'package' with standardised objects and, second, the local and exotic materialities that formed this 'package'. The paper will focus on the themes of original and skeuomorph (i.e. a copy of an object in a different material, usually more common, preserving its general aspect), by analysing the analogies between the shape and the materiality of objects.

The Varna prestige package

It can be argued that the best example to illustrate the value and prestige in the GKK emergent stratified society is the Varna necropolis (see Renfrew, 1986: 148-150). A detailed analysis of the graves and cenotaphs with rich inventory reveals sets of 'masculine' and 'feminine' objects, deposited separately [as in some cenotaphs] or together, such as in Grave 4 (fig. 1), which is a cenotaph containing both types of objects. In the case of cenotaphs the sex of the individuals can be inferred by comparing the anthropomorphic masks with the female miniature figurines, both of which exhibit common elements, such as a large number of earrings and metal pins fixed in the inferior lip, traits which are absent on male figurines or in tombs with a determined male skeleton. In the 'feminine' cenotaphs, like for example in Grave 3, the inventory consists of small ceramic vases decorated with graphite, a marble

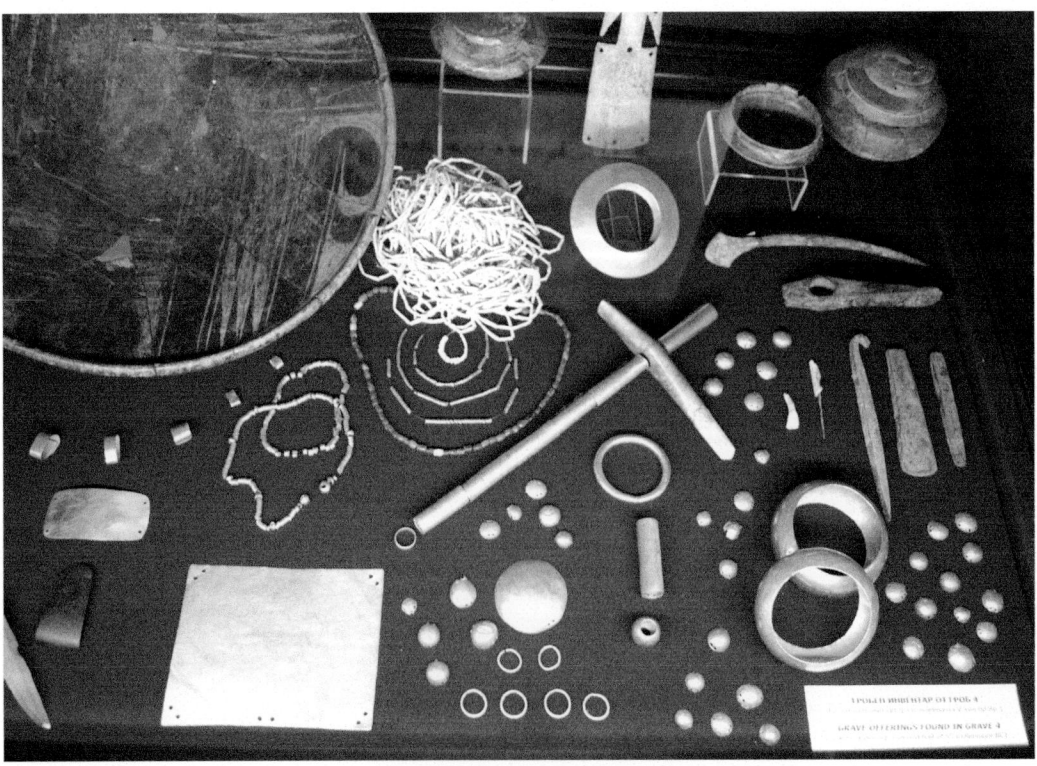

FIGURE 1. GRAVE 4 (CENOTAPH), VARNA CEMETERY (VARNA MUSEUM).

plate, a gold diadem, several small round convex perforated gold pendants that resemble the *Spondylus* shell (Gheorghiu, 2011a, 2012), gold earrings, gold lip-pins, *Dentalium* shell necklaces, an anthrozoomorphic figurine made of bone or marble, and short flint blades. Compared with this relatively wealthy inventory containing local and exotic materials, the men's burials are more spectacular because of the paraphernalia related to the body and costume. For example several objects of the Grave 43 inventory (which, compared to other burials could be considered to contain a complete set of 'masculine' prestige objects) (fig. 2), were designed to be worn on a headdress and on the costume (e.g. gold disks and plaques of different sizes), on the body (such as gold and Spondylus bracelets and different

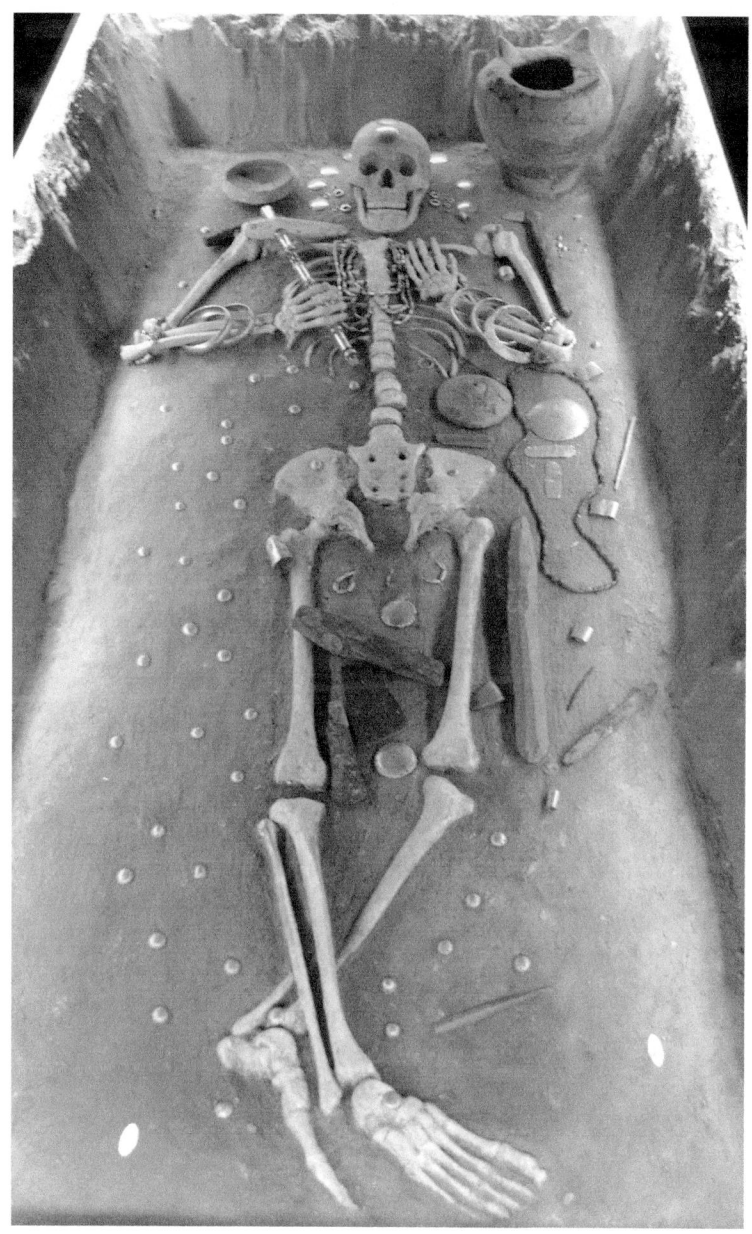

FIGURE 2. GRAVE 43 (VARNA MUSEUM).

necklaces made of gold and carnelian), or simply handled (such as different types of stone and copper axes, copper awls and long flint blades of various sizes) (fig. 3). The social visibility of the tools mentioned can be inferred from the sockets being covered with gold foil. In addition to the inventory mentioned the grave also contained a ceramic vase-support and a ceramic lid.

This set of objects, that form a sort of standardised prestige 'package', is composed of local and exotic insignia, namely objects to be displayed, whose main function was to quantify and exhibit the wealth or status of the owner (see Veblen, 1915). Similar sets of packages with insignia were found in the peripheral settlements north the Danube River, in Cucuteni-Tripolye tradition settlements, and, with important material interpretations, in the steppe population's graves from the West-Pontic region (Haheu and Kurciatov, 1993; Bicbaev, 2010).

Not all the packages discovered had a complete inventory; for example, in a dwelling at Sultana tell, north of the Danube River, only the gold objects of the prestige package were found in a ceramic

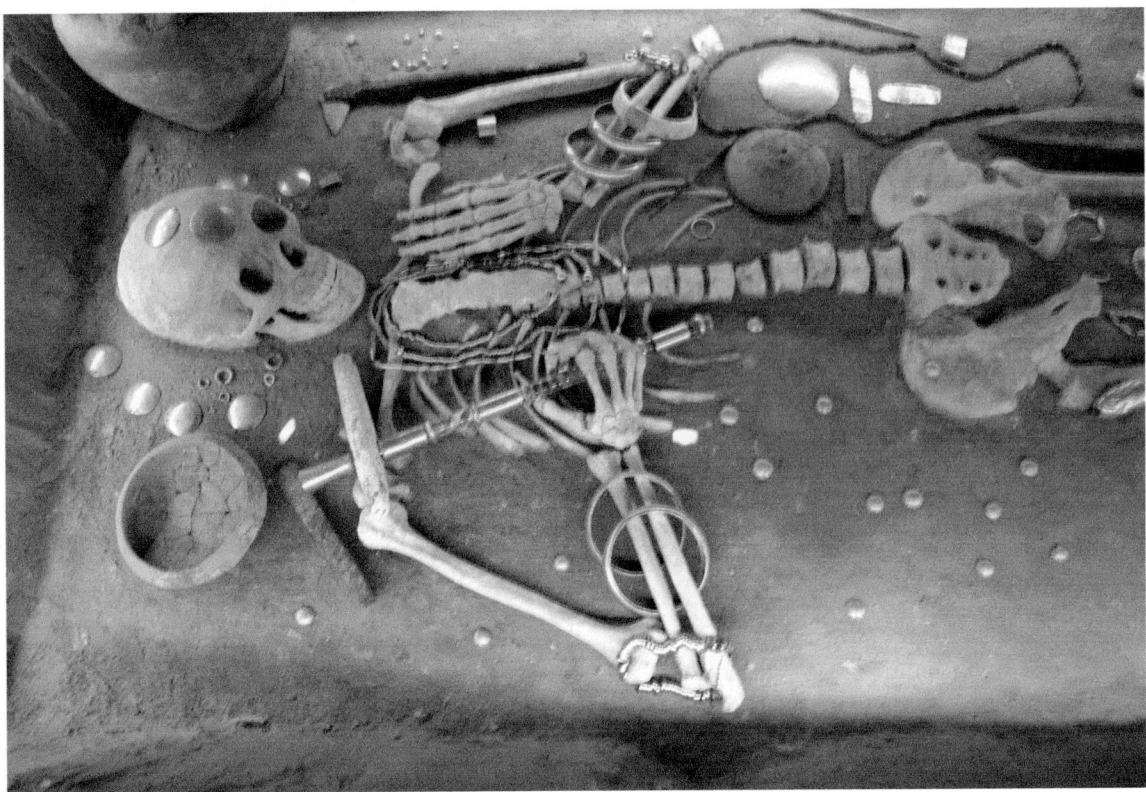

FIGURE 3. GRAVE 43, DETAIL. (VARNA MUSEUM).

vase. In the Cucuteni area, gold objects similar with the GKK ones (see Dumitrescu, 1961) suggest that parts of the prestige package circulated independently over long distances.

The materiality of the package: original and skeuomorph

The process of acquisition of materials played a key role in the creation of the prestige packages, combining the rarity of procurement (for exotica) with the difficulty of processing (for local metals). Since the exotica were the result of long trade routes, which implied a series of hazards, the practice of producing copies with local materials of value became a strategy to counterbalance an economic policy otherwise beset with high risks. For instance, the *Spondylus* shell, an item acquired from distant sources in the 5th millennium BC was copied in gold to produce small ritual objects with a central perforation, bracelets or necklaces (Gheorghiu, 2011a, 2012). Another item subject of skeuomorphism was the *Dentalium* shell. A good example could be Grave 4 where *Dentalium* necklaces are found together with gold necklaces made of thin gold foil rolled to look like the conical shape of the shell copied.

An additional example of the skeuomorphism of the exotic shells (Gheorghiu, 2011a, 2012) could be the prestige package from Sultana tell, situated 100 km north-west from Varna. The set of gold objects from Sultana, displaying perfect analogies with the 'masculine' and 'feminine' gold paraphernalia from Varna, consists of a series of earrings, bracelets and necklaces deposited in a ceramic vase in a dwelling (Hălcescu, 1995) (fig. 4).

As the distance from the centre increased, even the acquisition of the local materials from the Balkans became a problem; for the communities situated far from the centre of the GKK tradition, skeuomorphism represented the solution to preserve the shape and colour and, subsequently, the 'magic' of the prestige objects, which began to be copied in the local materials of value.

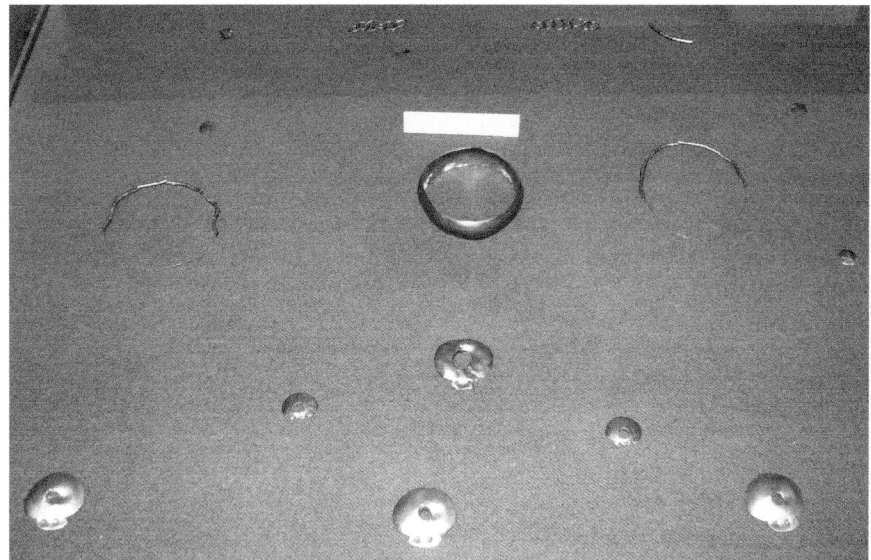

FIGURE 4. A REPLICA OF THE SULTANA PACKAGE (OLTENIȚA MUSEUM).

The materiality of the package as a technological index

In conjunction with the perception of the materiality of the prestige package as being a social instrument for establishing value and social status, another interpretation could be technological. Particularly for metals, the materiality of prestige objects functioned rhetorically as a metonymy referring to a whole occult technological process of acquisition and transformation of the ore. Consequently this kind of techno-shamanism (see Gheorghiu, 2011b: 81) generated by the technology of processing metals extended implicitly to the ownership and display of the metal objects, the package being perceived as a set of know-hows whose possession and display constituted an act of power and prestige.

The transmission of the package in Cucuteni-Tripolye could therefore be seen not only as the transmission of a political idea (i.e. the construction of social status with insignia), but also the knowledge of a set of technologies. In this respect, in Cucuteni-Tripolye, the skeuomorphic transfer in copper of some gold objects from Varna infers the existence of a local production (for local copper sources see Klochko *et al.*, 1999) generated by the exotica as shown by the packages from Cărbuna (Dergacev, 1998) and Brad (Ursachi, 1991).

The dispersion of the prestige package in the Cucuteni-Tripolye tradition

Some of the 444 copper objects from the Cărbuna hoard (found in a Precucuteni III ceramic vase) are a stronger stylistic interpretation of the V-shaped gold objects, of animal horns or of Spondylus valves from Gumelnița. This group of objects comprised two decorated ceramic vases, *Spondylus* bracelets, two copper axes (one of Vidra type), two hammer-axes (one made of marble and one made of schist), an awl, spiralled bracelets and pendants (Dergacev, 1998: 29; 43-44).

The Brad package (Cucuteni A3-A4 phases; Ursachi, 1991: 353) found in an *askos* vase (Gumelnita A2; Monah, 2003: 131) was formed by a set of copper objects (a Varna-Fâstâci axe with convex blade analogous with one from the Varna Grave 4, two spiralled bracelets, three convex discs with perforations, two necklaces made of small beads), two gold convex discs (analogous with those from the Graves 4 and 43), and a necklace made of deer teeth (fig. 5).

An additional example of interpretation of the prestige package, as well as the persistence of the cultural model after the disappearance of the GKK tradition could be the Horodnica II package

FIGURE 5. THE BRAD PACKAGE (ROMAN MUSEUM OF HISTORY) (FROM MANTU ET AL., 1997: 153, FIG. 151).

from Cucuteni phase B (Sulimirski, 1961: 96), that contained a massive double copper axe of Jászladány type, a copper awl, a copper dagger, a copper crown, and a necklace, made of cooper biconical beads (a skeuomorph of the *Spondylus* or cornelian necklaces), all stored in a painted vase (fig. 6).

Other cases of skeuomorphism are the Ariusd hoard (Cucuteni A2 phase) where prestige exotic materials like *Spondylus* were replaced with local materials like boar tusks (Sztáncsuj, 2005: 93, fig. 8), and local prestigious materials like deer canine teeth were replaced with skeuomorphs of bone and clay (Sztáncsuj, 2005: 92, fig. 7), and the Hăbășești hoard (Cucuteni A2 phase) where deer canine teeth were copied in bone (Beldiman and Sztancs, 2005: 108).

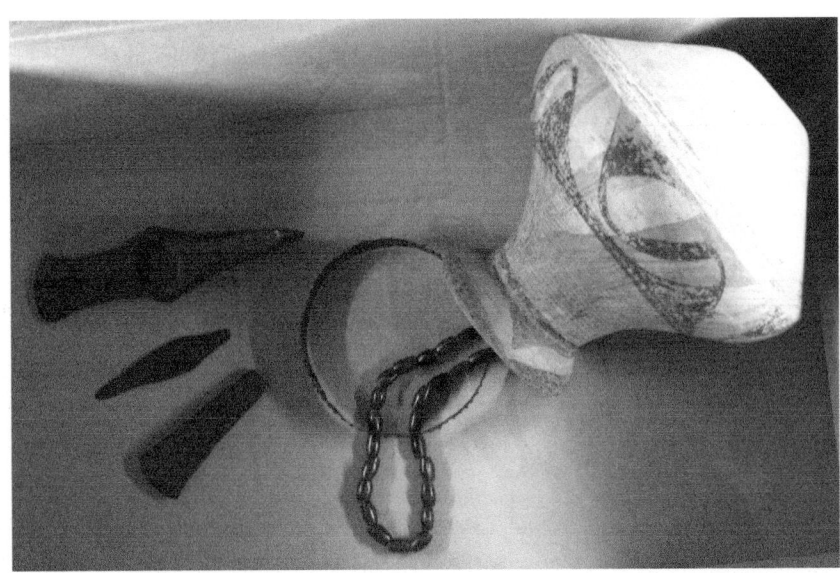

FIGURE 6. THE HORODNICA II PACKAGE (NATURAL HISTORY MUSEUM, VIENNA).

The spreading of the package to the steppe populations

A special form of skeuomorphism of the prestige package is to be found in the north Pontic tradition Suvorovo-Novodanilovka, as one can see from the inventory of the Giurgiulesti cemetery (Haheu and Kurciatov, 1993; Chapman, 2013: 326), particularly of the Grave 4. The special inventory of this tomb (Bicbaev, 2010: 218) comprises a set of objects analogous with the Varna package (a massive copper dagger, two gold spiraled rings, a shell bead), together with items that can be identified as being skeuomorphs (i.e. a deer antler spear point whose shaft was covered by three gold tubular fittings, and two round patterns made of coral beads as head decoration).

A special case of skeuomorphism is the half metre antler and wood spear shaft with small flint blades inserted on each side that creates the image of a long flint blade like the one from Grave 43 in Varna. The action of turning the objects of value from the agricultural societies into skeuomorphs could be observed in the use of the boar tusk plaques that resemble the fragmented *Spondylus* bracelets (Anthony, 2010: 38-9).

Prestige packages and rituality

All the prestige packages mentioned in the text also included a ceramic container that could indicate a ritual utilization together with the paraphernalia. The decoration of the costume and the body indicates a ritual performance (Turner, 1969; Bell, 1997: 159ff) and a complex social spectacle. All status insignia (Gheorghiu, 2011a: 20 ff.) from Varna that emerge at the periphery of the tradition or in neighbouring traditions as standardised assemblages are proof of the circulation of ideas and materials between the centre and the peripheries, and the formation of a common conception of value, prestige, technology and rituality (see also Monah, 2003).

Conclusion: Distance, materiality and the persistence of the cultural model. The role of skeuomorphs in creative peripheries

The Chalcolithic prestige package discussed represented a complex pattern of material and temporal compositions and relationships between objects (see Gheorghiu, 2001), of relationships between objects and people (see Hodder, 2012; Latour, 2005; Renfrew, 2004; Appadurai, 1986), and between society and the images created by people (see Bradley, 2009).

The spatial diffusion and temporal persistence of the prestige package could be proof it represented an important cultural model, with a noticeable theatrical character. It is known that public events, especially ritual actions, comprise an amount of spectacle (Inomata and Coben, 2006: 16), and in this respect the package can be perceived as a set of objects of display, a set of garments of a theatrical costume, where skeuomorphs accentuated the theatrical character of the owner. The utilization of skeuomorphs represented an efficient and creative solution to solve the problems of acquisition of rare exotica brought through long distance trade, also playing a theatrical role of artificial decoration that replaced the original materials.

Skeuomorphs belonged to a visual system of communication that represented a creative solution by the peripheries to assimilate a cultural model created by a centre. The skeuomorphs of *Spondylus*, *Dentalium*, gold objects, or long flint blades allowed the exhibition of an artificial insignia in the place of the original, that facilitated the performance, both corporeal and political (see Hodder, 2006: 85) of the display of the prestige within the community in the theatre of power even at the small scale of the prehistoric villages (see Hodder, 2006: 82 ff.).

In the Cucuteni-Tripolye tradition even if the materiality of the package was sometimes changed, the image produced by the new materialities of the skeuomorphs tried to reproduce the original model, and, consequently the original ritual action.

The change in the materiality that occurs in the east European Chalcolithic, by the deliberate replacement of the organic exotica with metal objects (see Jovanović, 1996: 31; Lichardus, 1991), infers the existence of a process analogous to the one from Late Neolithic tradition in Central Europe, considering 'copper as a new prestige raw material was accessible only to a narrow group and it was able to take over the role which *Spondylus* ornaments were more and more unable to play' (Siklosi and Csengeri, 2011: 57; see also Anthony, 2010: 38).

All the pieces of the prestige package represent a standardised form of value and prestige (and probably of ritual actions) indicating the existence of an inter-settlement, long distance transmission of ideas and materials and the existence of a trans-cultural homogeneity extended over several centuries.

Acknowledgements

The author wishes to thank Professors Marie Besse and Jean Guilaine for the kind invitation to contribute to the present volume. Many thanks also to M. Bogdan Căpruciu for proof-reading the text.

The documentation in Varna was possible due to a PN II IDEI research grant (Time Maps).

Photos 1-4 and 6 by the author.

References

Anthony, D. W., 2010 – The Rise and fall of Old Europe, pp. 29-57. Pp. 163-177. In: Anthony, D. W.; Chi, J. Y. (eds.) The Lost World of Old Europe. The Danube Valley, 5000-3500 BC. Princeton and Oxford: Princeton University Press.

Appadurai, A. (ed.), 1986 – The Social life of things. Commodities in cultural perspective. Cambridge: Cambridge University Press.

Beldiman, C.; Sztancs, D. M., 2005 – Les objets de parure en matières dures animales de la culture Cucuteni: Le dépot de Hăbășești, Dep. de Iași, pp. 107-115. In: Dumitroaia, Gh.; Chapman, J.; Weller, O.; Preoteasa, C.; Munteanu, R.; Nicola, D.; Monah, D. (eds.) Cucuteni. 120 ans des recherches. Le temps du bilan/120 Years of Research. Time to sum up. Bibliotheca Memoriae Antiquitatis, XVI, Piatra Neamț.

Bell, C., 1997 – Ritual: Perspectives and dimensions, Oxford: Oxford University Press.

Bicbaev, V., 2010 – The Copper Age cemetery of Giurgiulesti, pp. 2013-225. In: Anthony, D.; Chi, J. I. (eds.) The Lost World of Old Europe. The Danube Valley, 5000-3500 BC. Princeton and Oxford: Princeton University Press.

Bradley, R., 2009 – Image and audience. Rethinking prehistoric art. Oxford: Oxford University Press.

Chapman, J., 2013 – From Varna to Brittany via Csőszhalom – Was There a 'Varna Effect?', pp. 323-336. In: Anders, A.; Kalla, G.; Kiss, V.; Kulcsar, G.; Szabo, G. V. (eds.) Moments in time. Budapest: L'Harmattan.

Dergacev, V. A., 1998 – Karbunski clad. Chishinev.

Dragomir, I., 1983 – Eneoliticul din sud-estul României. Aspectul Stoicani-Aldeni. Bucharest: Academiei RSR.

Dumitrescu, H., 1961 – Connections between the Cucuteni-Tripolye cultural complex and the neighbouring Eneolithic cultures in the light of the utilization of golden pendants. Dacia N. S. V: 69-93.

Frînculeasa, A., 2011 – Seciu – Judetul Prahova. Un sit din epoca neo-eneolitica in nordul Munteniei. Bucharest: Oscar Print.

Gheorghiu, D., 2001 – Material, Virtual and Temporal Compositions: On the Relationship between Objects. British Archaeological Reports. Oxford: Archaeopress.

Gheorghiu, D., 2011a – Insignia of exotica: Skeuomorphs of Mediterranean shells in Chalcolithic south eastern Europe, pp. 13-25. In: Vianello, A. (ed.) Exotica in the Mediterranean. Oxford: Oxbow.

Gheorghiu, D., 2011b – Working with Agni: The Phenomenological Experience of a Technological Ritual, pp. 71-88. In: Gheorghiu, D. (ed.), Archaeology Experiences Spirituality?, Newcastle upon Tyne: Cambridge Scholars Publishing.

Gheorghiu, D., 2012 – 'Skeuomorphs' on the rhetoric of material in the Gumelnita tradition. Documenta Praehistorica XXXIX, pp. 287-294.

Guilaine, J., ed., 2004 – Aux marges des grands foyers du Néolithique. Périphéries débitrices ou créatrices? Paris: Errance.

Haheu, V.; Kurciatov, S., 1993 – Cimitirul plan eneolitic de langa satul Giurgiulesti, Revista arheologica I, pp. 101-114.

Hălcescu, C., 1995 – Tezaurul de la Sultana, Cultura si civilizatie XIII-XIV, pp. 11-15.

Higham, T.; Chapman, J.; Slavchev, V.; Gaydarska, B.; Honch, N.; Yordanov, Y.; Dimitrova, B., 2007 – New perspectives on the Varna cemetery (Bulgaria) – AMS dates and social implications. Antiquity 81: 640-651.

Hodder, I., 2012 – Entangled. An Archaeology of the relationship between humans and things, Malden, Oxford: Wiley-Blackwell.

Hodder, I., 2006 – The Spectacle of daily performance at Çatahöyük, pp. 81-102. In: Inomata, T.; Coben, L. S. (eds.) Archaeology of performance. Theaters of power, community and politics. Lanham, New York, Toronto, Oxford: Altamira Press.

Inomata, T.; Coben, L. S., 2006 – Ouverture: An invitation to the archaeological theater, pp. 11-46. In: Inomata, T.; Coben, L. S. (eds.) Archaeology of performance. Theaters of power, community and politics, Lanham, New York, Toronto, Oxford: Altamira Press.

Ivanov, I., 1988 – Die Ausgrabungen des Gräberfeldes von Varna.pp. 49-66. In: Fol, A.; Lichardus, J. (eds.), Macht, Herrshaft,und Gold. Saarbrücken, Moderne Galerie des Saarland Museums.

Ivanov, I.; Avramova, M., 2000 – Varna necropolis: The Dawn of European civilization. Sofia: Agatho.

Jovanović, J., 1996 – Eneolithic gold pendants in South-East Europe: their meaning and their chronology. pp. 31-36. In: Kovács, T. (ed.) Studien zur Mettalindustrie im Karpatenbecken und den banachbarten Regionen: Festschrift für Amália Mozsolics zum 85. Geburtstag, Budapest: Magyar Nemzeti Múzeum.

Klochko, V. I.; Manichev, V. L.; Kvasnitsa, V. N.; Kozak, S. A.; Demchenko, L. V.; Sohkatski, M. P., 1999 – Issue concerning the Tripolye metallurgy and the virgin copper of Volhynia, Baltic Pontic Studies 9, pp. 168-186.

Latour, B., 2005 – Reassembling the social: An Introduction to Actor-Network theory. Oxford: Oxford University Press.

Lichardus, J., 1991 – Das Gräberfeld von Varna im Rahmen des Totenrituals des Kodžaderm-Gumelnița-Karanovo VI-Komplexes. Pp. 167-194. In: Lichardus, J. (ed.) Die Kupferzeit als historische Epoche: Symposium Saarbrücken und Otzenhausen 6.-13. 11. 1988. Bonn: Dr. Rudolf Habelt GmbH.

Mantu, C-M.; Dumitroaia, Gh.; Tsaravopoulos, A. eds., 1997 – Cucuteni – The last great Chalcolithic civilization of Europe [cat.], Thessaloniki: Athena Publishing and Printing House.

Monah, D., 2003 – Quelques réflections sur les trésors de la culture Cucuteni, Studia Antiqua et Archaeologica IX, pp. 129-140.

Pernicka, E.; Anthony, D. W., 2010 – The invention of copper metallurgy and Copper Age of Old Europe. pp. 163-177. In: Anthony, D. W.; Chi, J. Y. (eds.) The Lost World of Old Europe. The Danube Valley, 5000-3500 BC. Princeton and Oxford: Princeton University Press.

Popovici, D. N., 2010 – Copper Age traditions north of the Danube River, pp. 91-111. In: Anthony, D.; Chi, J. I. (eds.) The Lost World of Old Europe. The Danube Valley, 5000-3500 BC, Princeton and Oxford: Princeton University Press.

Renfrew, C., 1986 – Varna and the emergence of wealth, pp. 141-168. In: Appadurai, A. (ed.) The Social life of things. Cambridge: Cambridge University Press.

Renfrew, C., 2004 – Towards a theory of material engagement, pp. 23-32. In: Demarrais, E.; Gosden, C.; Renfrew, C. (eds.) Rethinking materiality. Cambridge: McDonald Archaeological Institute.

Rusev, R.; Slavcev, V.; Marinov, G.; Boyadjev, I., 2010 – Varna – Praistoricesky tzentr na metaloobrabotkata. Varna: Dangrafik.

Séfériades, M., 2010 – Spondylus and long-distance trade in prehistoric Europe, pp. 179-190. In: Anthony, D.; Chi, J. I., (eds.) The Lost World of Old Europe. The Danube Valley, 5000-3500 BC. Princeton and Oxford: Princeton University Press.

Siklósi, Z.; Csengeri, P., 2011 – Reconsideration of Spondylus usage in the Middle and Late Neolithic of the Carpathian basin, pp. 47-62. In: Ifantidis, F.; Nikolaidou, M. (eds.) Spondylus in Prehistory: New Data and Approaches – Contributions to the Archaeology of Shell Technologies. BAR International Series 2216, Oxford: Archaeopress.

Sulimirski, T., 1961 – Copper hoard from Horodnica on the Dniester, Mitteilungen des Anthropologischen Gesellschaft in Wien XCI: 91-96.

Sztáncsuj, S. J., 2005 – The Early Copper Age hoard from Ariuşd (Erösd), pp. 85-105. In: Dumitroaia, G. H.; Chapman, J.; Weller, O.; Preoteasa, C.; Munteanu, R.; Nicola, D.; Monah, D. (eds.) Cucuteni. 120 ans des recherches. Le temps du bilan/120 Years of Research. Time to sum up. Piatra Neamţ: Bibliotheca Memoriae Antiquitatis XVI.

Todorova, H., 1982 – Kupferzeitliche Siedlungen in Nordostbulgarien. München: C. H. Beck.

Turner, V., 1969 – The Ritual process: structure and anti-structure. Ithaca, NY: Cornell University Press.

Ursachi, V., 1991 – Le dépot d'objets de parure enéolithique de Brad, Le paléolithique et le néolithique de la Roumanie en contexte européen. Biblitheca Archaeologica Iassiensis IV: 335-386.

Veblen, T., 1915 – The Theory of the leisure class. An economic study of institutions. New York: The Macmillan Company; London: Macmillan and Co. Ltd.

Exchange and interaction: the Iberian Mediterranean between the VI and III millennia cal BC

Teresa Orozco Köhler and Joan Bernabeu Aubán
Universitat de València

Abstract

Exchange and interaction processes show a development from the Early Neolithic. Through various examples chosen in the Mediterranean Iberia we try to draw the dynamic nature of these processes, in a general way. Not only materials that can be considered exotic, or valuable, circulating over long distances; also some domestic objects show interesting distributions. In other cases, we can observe the flow of information, or technology. Some examples illustrate the importance of social motivations have for human groups involved in them.

Key words: *Exchange networks, Neolithic, Mediterranean Iberia*

Résumé

Les processus d'échange et d'interaction montrent une évolution dès le Néolithique ancien. Grâce à divers exemples choisis dans la Méditerranée ibérique, nous essayons de souligner la nature dynamique de ces processus. D'une part, nous discuterons non seulement des matériaux qui peuvent être considérées comme exotiques, ou de valeur, circulant sur de longues distances, mais aussi de quelques objets domestiques qui montrent des distributions intéressantes. D'autre part, nous observerons la circulation de l'information ou de la technologie. Quelques exemples proposent d'illustrer l'importance des motivations sociales des groupes humains qui y participent.

Mots clés: *Réseaux d'échanges, Néolithique, Méditerranéenne Ibérique*

Introduction

Material objects that make it possible to trace circulation as either raw materials or manufactured goods provide an excellent focus for analysing exchange during prehistory. From the VI millennium cal BC we see the processes of interaction between groups undergo constant development with chronological and regional variations. These processes can be seen through the presence of various different materials: flint, obsidian, and a number of adornments and symbolic objects. The Neolithic sees a real revolution take place compared to previous stages, with a greater intensity of contacts and even what some would call an 'exchange economy'. In recent years there have been numerous investigations that from different viewpoints have enabled us to increase our knowledge of this phenomenon. Two areas in which there has been significant development are the creation and fine tuning of new archaeometric methods, which have brought about improvements in the characterization of raw materials, and the investigation of the places in which these materials were extracted (mines and quarries).

However, advances have not been made to the same degree as regards the theoretical frameworks that enable us to make all these phenomena understandable. This brief text aims to focus the attention not only on the ways in which materials circulated or the methodologies that allow them to be identified, but also on the social networks involved and the need to apply new approaches to studying them.

The examples chosen are from an Iberian context, mainly affecting the Mediterranean area. The distributions and their geographical framework are briefly presented without going into a detailed description of the collections and sites. A general outline of the time scales of the various collections has also been given so as to highlight the dynamic character of the distributions over a lengthy chronological axis. In the final section we emphasize the need for additional tools for interpretations.

New data on Iberian Late Prehistory: characterization of lithic materials, mines and quarries

There have been notable advances in the archaeometry of lithic materials in recent years. The quantity and quality of the work carried out in this area have increased exponentially in the application of both optical methods and various chemical analyses (Shackley, 2008). The constant exploration of minimally invasive techniques to use in the determination of archaeological pieces is reflected in the numerous papers that appear in the bibliography. The last few years have seen an exhaustive search for non-destructive analysis techniques that are quick and portable. In this respect we should cite the work carried out on a European scale by the JADE project on the characterization and distribution of alpine axes on the continent during the V and IV millennia cal BC, which has achieved excellent results using spectroradiometry on lithic materials (Petrequin *et al.*, 2012).

Concern for the determination and characterization of source areas and the quality of the data from field sampling is to be found among many Iberian researchers, who, aware of the effort involved, are setting up initiatives for sharing institutional lithic collections.

Another area in which there has been a notable increase in knowledge in recent years is the documentation and study of sites specializing in production. The mining of the subsoil by way of shafts and galleries in prehistoric times has traditionally received a great deal of attention, possibly because of the spectacular nature of the structures. However, other types of site such as open-cast workings (quarries) and also workshop areas provide data of enormous interest in helping us to discover more about operating systems and the transformation of particular resources.

Data on the Iberian Peninsula have grown substantially in recent times. We will take a look at two examples located in the central and south-western areas: the discovery and characterization of mineworks in Casa Montero (Madrid) and Pico Centeno (Huelva).

The findings in the Neolithic mine in Casa Montero show around 4000 vertical pits, in some cases as much as 10 metres deep, used for obtaining flint (fig. 1). The nodules extracted were mainly used to make blades. A huge amount of archaeological remains and waste materials are to be found both outside the pits and filling part of the structures, enabling us to get a fairly clear idea of the tasks carried out there (Capote *et al.*, 2008; Diaz-del-Rio *et al.*, 2008; 2010). Such a large number of localized structures might at first sight be interpreted as the result of a supply model consisting of short episodes dedicated to mining over a long time scale. However, radiocarbon dating indicates that the work and extraction activities at the mine took place over a short period of time, around a century (Diaz-del-Rio and Consuegra, 2011).

The pits in this flint mine do not cut through each other and present a fairly homogeneous morphology. Sometimes the walls show evidence of being adapted to make working conditions easier. The pit-mining system in Casa Montero is considered safe and efficient. The structures would have been filled in immediately with the remains and waste materials from excavation and lithic reduction, thereby preventing accidents. The reason Neolithic groups chose this spot to obtain flint is considered to be the result of a combination of factors: geological, technological, locational and especially social (Diaz-del-Rio and Consuegra, 2011).

The other prehistoric mineworkings on the Peninsula that are being recently studied in detail – the variscite mines in Pico Centeno – are of a different nature. The lithic material here was mined for manufacturing beads at certain stages of recent prehistory (Linares and Odriozola, 2011). We have a great deal of information about the variscite mines in Gavà (Barcelona), but the case of Pico Centeno shows certain differences that may possibly be related to the nature of the geological substrate and also the area's topography. In this case the mineral was extracted using the open-cast method (fig. 1). These are typically trench mines that follow the vein of ore. Researchers have found large quantities of debitage and abundant mining tools in the area. The evidence suggests that the Pico Centeno mines were used for the production of variscite nodules and pre-formed blanks. There is no evidence of

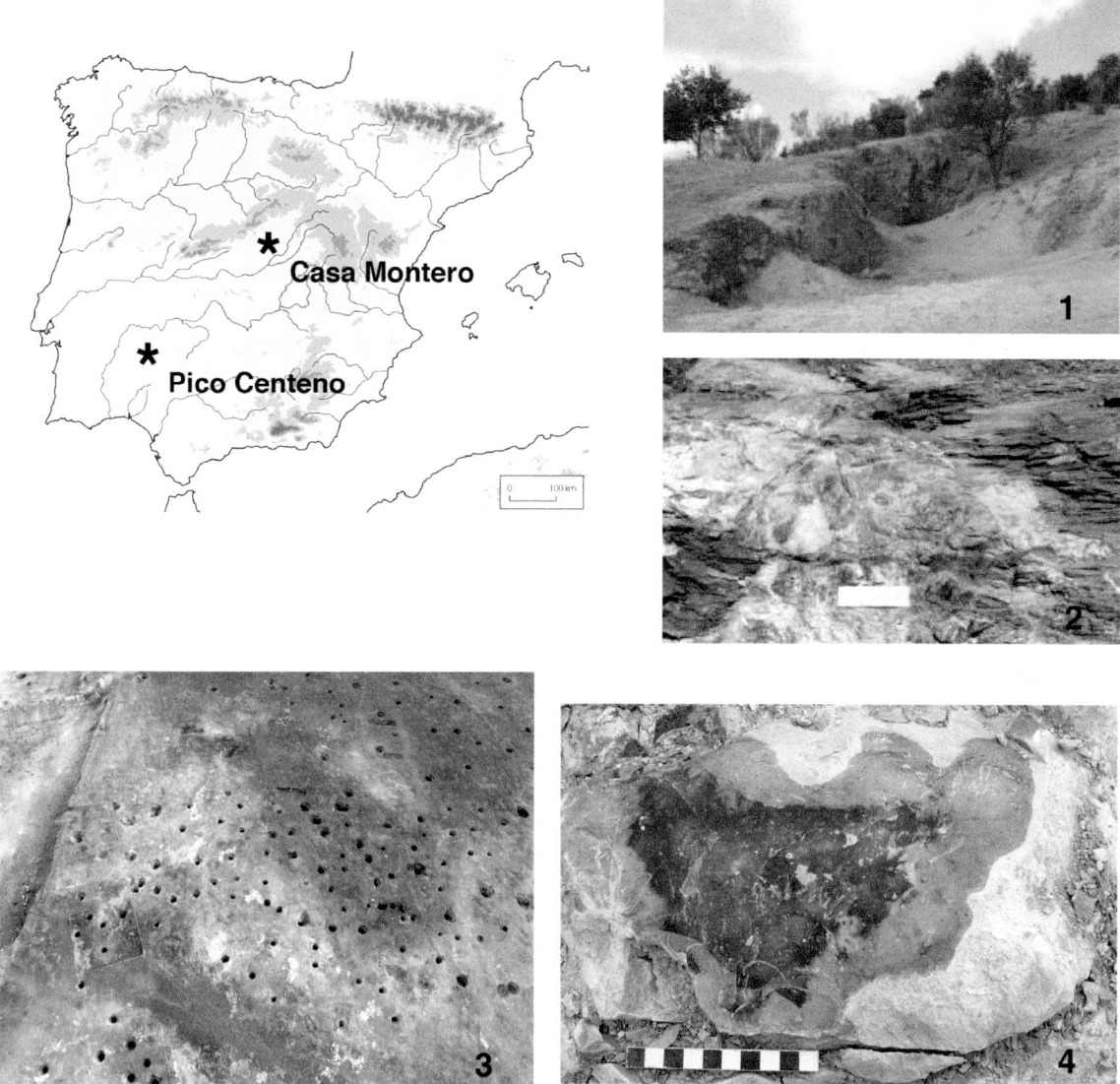

FIGURE 1. SOME OF THE RAW MATERIAL SOURCES RECENTLY DISCOVERED IN IBERIA: PICO CENTENO (VARISCITE), CASA MONTERO (FLINT). 1) A TRENCH MINE IN PICO CENTENO; 2) FILONIAN VARISCITE MINERALIZATION (AFTER LINARES AND ODRIOZOLA, 2011); 3) AERIAL VUE OF CASA MONTERO NEOLITHIC MINE; 4) FLINT NODULE OF CASA MONTERO (HTTP://WWW.CASAMONTERO.ORG/REC_ALBUM.HTML).

further reduction activity, which suggests that the production processes were continued elsewhere (Odriozola *et al.*, 2010). The detailed characterization of this mineral is widening the debate on the origins, production and distribution of beads, establishing the areas of origin by comparison with geological samples from different areas of the Peninsula (Odriozola *et al.*, 2013).

Other archaeological materials (ivory, obsidian) have been the subject of recent research on the Iberian Peninsula, providing us with more detailed information about their production and distribution over the territory.

Distribution of materials: some examples from Mediterranean Spain

From the start of the Neolithic there is evidence of a substantial increase in the volume of products and raw materials circulating over long distances, while increased mining activities are documented

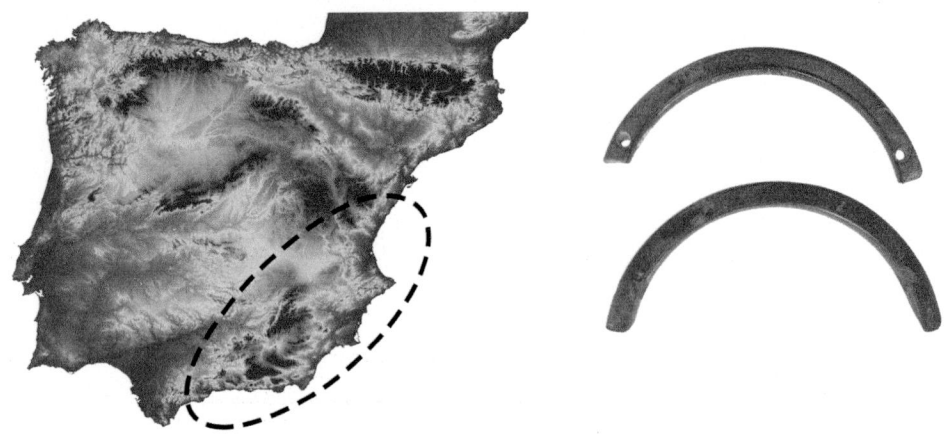

FIGURE 2. AREA OF SCHIST BRACELETES DISTRIBUTION IN THE MEDITERRANEAN IBERIA ALONG THE EARLY NEOLITHIC.

as occurring at the same time. The approach usually used to study the distribution and/or circulation of lithic tools and raw materials and similar elements has been refined over the years. Nowadays a combined analysis is carried out that includes mineralogical characterization, localization of source areas and a study of the work processes, product distribution and context of use, while at the same time taking into account that the social context in which these objects belong (domestic, funerary, symbolic, etc.) may affect their circulation.

On the Iberian Peninsula we see evidence of the circulation of different products and techniques throughout the time scale considered. Using different maps we can position various examples without going into an exhaustive analysis of the distributions. Overall we see the outline of a number of circuits of different sizes whereby objects and materials of different natures and characteristics are circulated. Their distribution in many cases overlaps in space and time. Nevertheless, a mere glance is enough for us to discern the dynamic of contacts and circulation. As we will see, it is not only materials that might be considered exotic or valuable that circulate over long distances; domestic objects also show some interesting distributions.

In the early stages of the Neolithic, schist bracelets are present in the records that extend across the southern and central areas of the Iberian Mediterranean (fig. 2). These are elements of personal adornment that show a certain homogeneity in their dimensions and nature, and possibly in their meaning and social value. Their distribution over the territory is more limited than that of other bracelets made of other materials (shell, limestone). The idea that these objects were exceptional has already been noted in previous papers (Bernabeu *et al.*, 2006) on the basis of their appearance at certain sites. They have been recovered not only from caves but also from a number of open-air sites, although it is interesting that they do not appear in funerary contexts. Although it has not yet been possible to establish their chronology with any accuracy, it seems to be limited to those stages in which cardial pottery was developed in Iberia.

Based on excavation work on the site at Cabecicos Negros (Almería), the schist bracelet production process has been documented in detail (Goñi *et al.*, 1999, 2000). The considerable number of pieces in the process of being made has enabled the various stages and procedures of the *chaîne opératoire* to be identified (fig. 3). The process is designed to produce highly standardized pieces. Cabecicos Negros is therefore seen as a workshop location that processed the lithic material that outcropped in the area. However, as yet we have no detailed characterization analysis enabling us to specify the distribution and circulation of these objects.

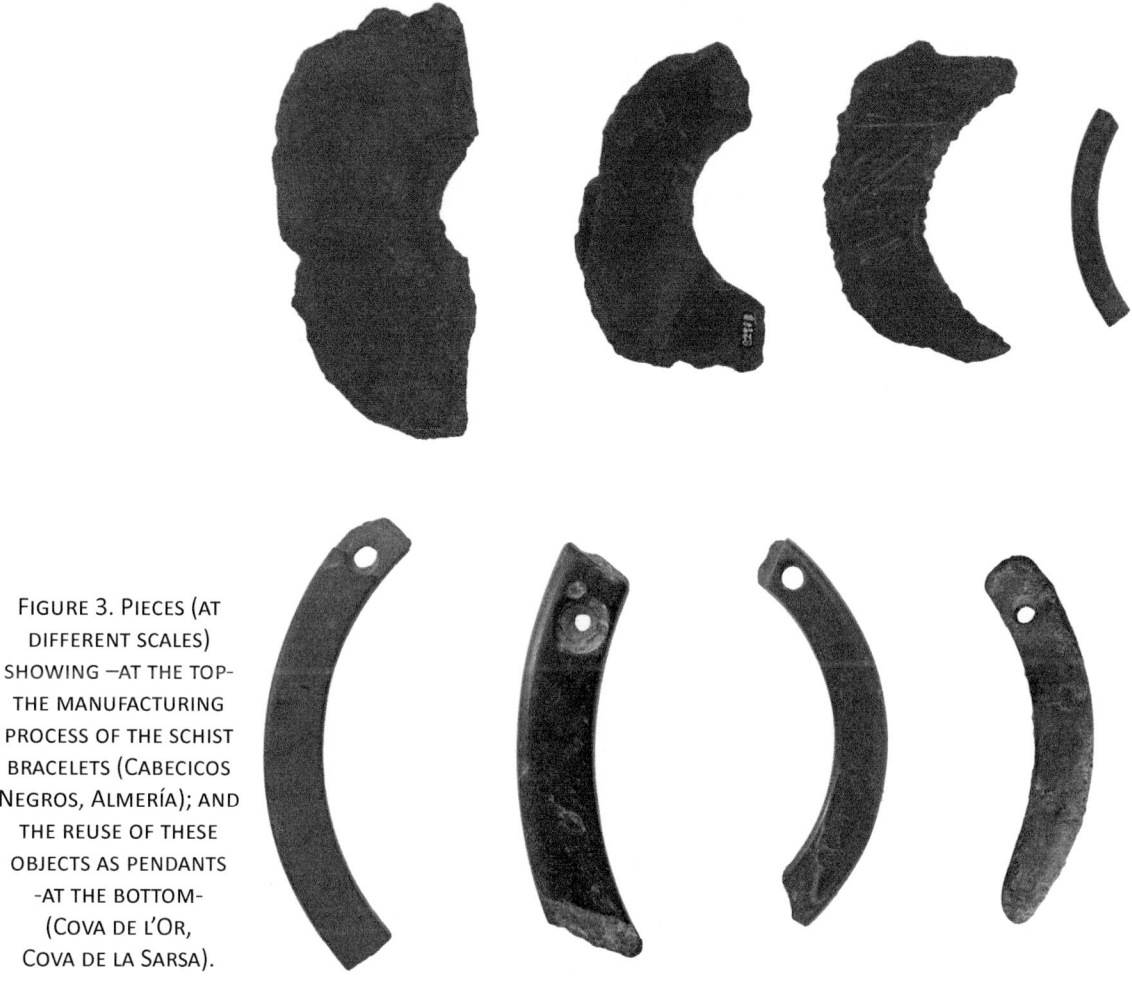

FIGURE 3. PIECES (AT DIFFERENT SCALES) SHOWING –AT THE TOP– THE MANUFACTURING PROCESS OF THE SCHIST BRACELETS (CABECICOS NEGROS, ALMERÍA); AND THE REUSE OF THESE OBJECTS AS PENDANTS –AT THE BOTTOM– (COVA DE L'OR, COVA DE LA SARSA).

Another example is provided by the circulation of polished amphibolite axes. These pieces are made of stone originating in the south-east of the Peninsula and their distribution stretches as far as the central Mediterranean area (fig. 4). Generally speaking this geographical framework is similar to the one considered in the above case of schist bracelets.

In this case we know the product and the contexts in which it was used, but the source area has only been identified in general terms and no workshop has been located, no site where it was made (Orozco, 2000). The presence of polished amphibolite pieces in the Valencia area corresponds to a cultural choice. In this case the qualitative features of this lithic material (toughness, visual aspect) are similar to those of tools made of materials that outcrop in the surrounding region (dolerites) and there are no typological differences between instruments made with either type of lithic material.

Although this distribution starts to appear in the early stages of the Neolithic, it would be from the IV and III millennia cal BC that the flow of materials would be strongly represented on archaeological sites, emphasizing the consolidation of this arrangement of circulation and contacts. The value of the polished axes that circulate does not stem from the exceptional nature or uniqueness of the lithic material. Indeed, bearing in mind the worn areas and breaks that tend to be present, their use is not considered to have been uncommon. Neither can possession of them be considered restricted since they appear in both habitat and funerary contexts. This means we can assume that a good proportion of the community had access to these goods (Orozco, 2011).

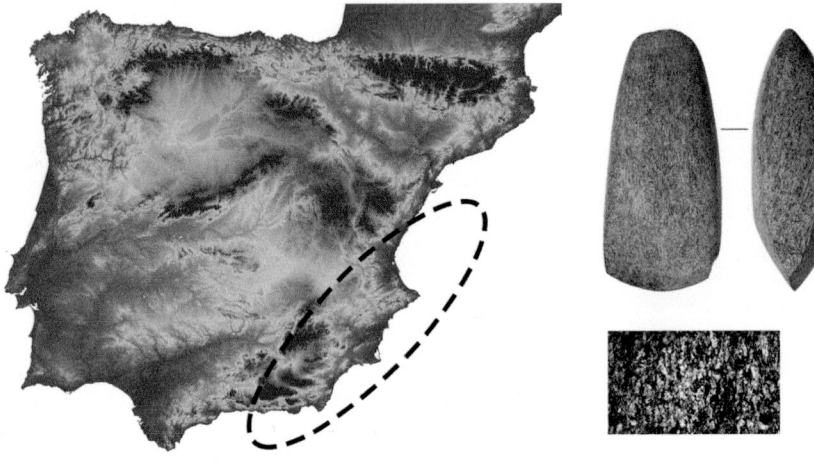

FIGURE 4. THE DISTRIBUTION OF AMPHIBOLITE STONE AXES FROM THE SOURCE AREA IN THE SOUTHEAST ARRIVES TO THE CENTRAL MEDITERRANEAN. THESE POLISHED TOOLS WILL BE IN CIRCULATION UNTIL THE BEGINNING OF BRONZE AGE.

Various materials enable us to visualize relationships that may possibly be of a different nature. Throughout the Peninsula we can find signs of distributions that point to the circulation and spread of technological characteristics. Studies made of the traces of use involving lithic pieces such as sickles that make up a typically Neolithic set of instruments present a suggestive pattern (fig. 5). These studies indicate that certain technical features of the tools enable interesting regional differences to be established. Taking into account a chronological framework stretching from the VI to the IV millennia cal BC (Gibaja *et al.*, 2010), in general terms it can be seen that sickles from the south of the Peninsula are clearly different to those from other geographical areas (central and north-east). They are made up of small stone flakes or fragments that are inserted diagonally into the handle. Sickles in the north-east are usually made with blades inserted parallel to the handle. Other types of sickle with their own particular characteristics have also been documented. It should be stressed that these differences in fitting the handle are not related to working with different types of plant species. Researchers believe that the different ways in which tools were made could be due to different cultural traditions, an aspect that work in progress will expand upon.

Other examples in the Mediterranean area of the Peninsula draw our attention to a particular aspect: the direction in which materials circulated. It we look at the distribution of polished hornfels axes, whose source area is in the north-east, despite the fact that the volume of pieces reaching the central Mediterranean area (fig. 6) is low, what is noticeable is the directionality of the flow of pieces, showing points of contact between these areas. This circulation begins in the mid-V millennium and

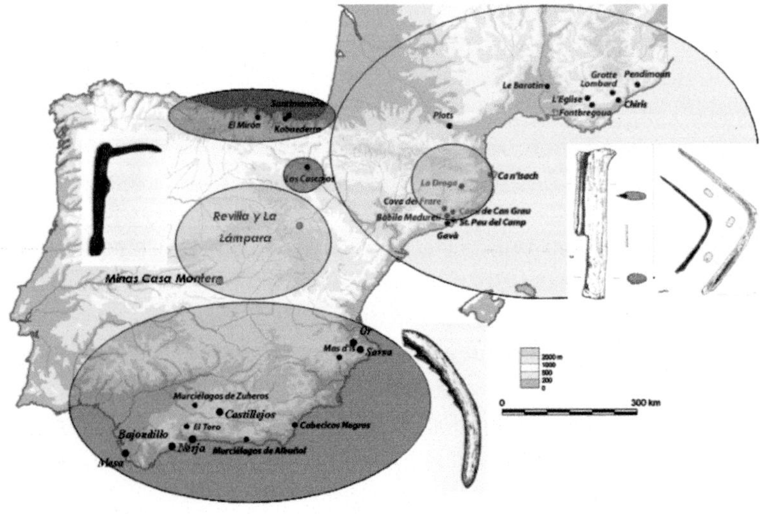

FIGURE 5. SOME TECHNOLOGICAL FEATURES ALLOW TO DELIMIT CLEARLY REGIONAL DIFFERENCES IN IBERIA. IN THIS CASE THE DIFFERENT TYPES OF SICKLES ARE NOT RELATED TO USE ON DIFFERENT PLANTS (AFTER GIBAJA *ET AL.*, 2010).

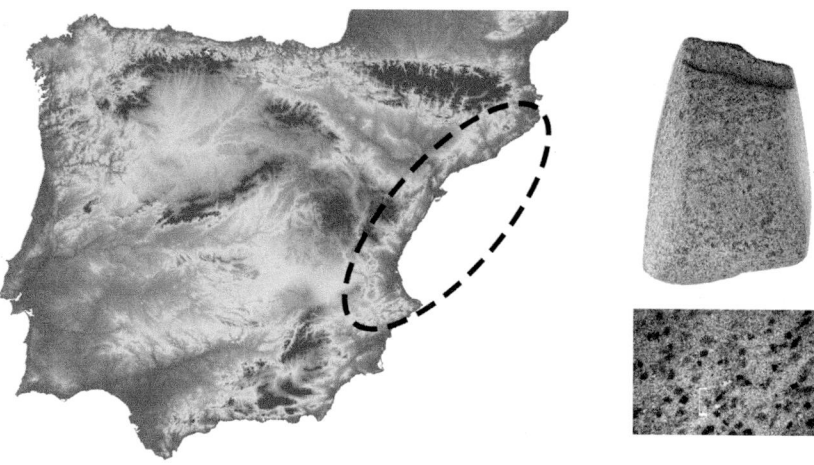

FIGURE 6. HORNFELS AXES AND ADZES HAVE THEIR SOURCE AREA IN THE NORTHEAST MEDITERRANEAN. THESE TOOLS CIRCULATE SOUTHWARD IN SMALL AMOUNTS.

continues through to the III millennium cal BC, although it appears always to be limited to just a few objects (Orozco, 2000; 2011).

The north-east of the Peninsula is not only the starting point from which some objects circulate. This territory also receives materials of enough singularity to enable their area of origin to be traced. We refer to certain elements of obsidian, one of the materials traditionally looked at in studies on exchange in the Neolithic Mediterranean. Analysis has recently been carried out on pieces recovered from a number of Catalan sites, the only ones documented in the whole of the Iberian Peninsula. This has made it possible to establish not only their chronological framework (late-V/early-IV millennia cal BC) but also their origin as being linked to source areas located on the island of Sardinia (Terradas *et al.*, 2014). These artefacts reflect the maximum geographical distance reached by the distribution of Sardinian obsidian (at least 1200 km), assuming it came via Provence (France) from where it could be redistributed along with other products. Although a good number of these elements have been recovered in funerary contexts, forming part of the collection of objects buried with the dead in individual tombs, recent investigations also show that they appear in domestic contexts (Gibaja *et al.*, 2014).

Generally speaking we find the outline of differently-scaled circuits via which objects and materials of different features and characteristics circulate. There are many elements from the archaeological record that also enable us to see patterns of distribution that may reflect their circulation over a particular territory.

As a final example we will cite a distribution of a different nature that we can trace across Peninsula territory. It involves the expression of a symbol, an eye-shaped motif, which is represented on mobile elements of various materials (limestone, slate, bone, ivory...) as well as in figures painted and/or etched on cave walls and megalithic monuments. Numerous occurrences can be found within an extensive geographical framework stretching from the south-west to the central and eastern Peninsula, throughout the IV and III millennia cal BC and in various contexts (funerary, domestic). Although we are still a long way from knowing the specific function of the different objects on which this iconography is shown, we can see that they carry a similar composition design (eyes and facial tattoo) adapted to different media, in which a representation of the same theme or symbol can be appreciated. A detailed analysis of these symbolic artefacts enables us to recognize regional variants and styles, which could be understood as territorial identity markers (Hurtado, 2008). In our case it would be interesting to highlight the common nexus between these figurative forms, showing the adoption and adaptation of an idea or theme that is shared between groups – and interpreted with variants – and its circulation over a wide territory, revealing relations and connections between different areas (fig. 7).

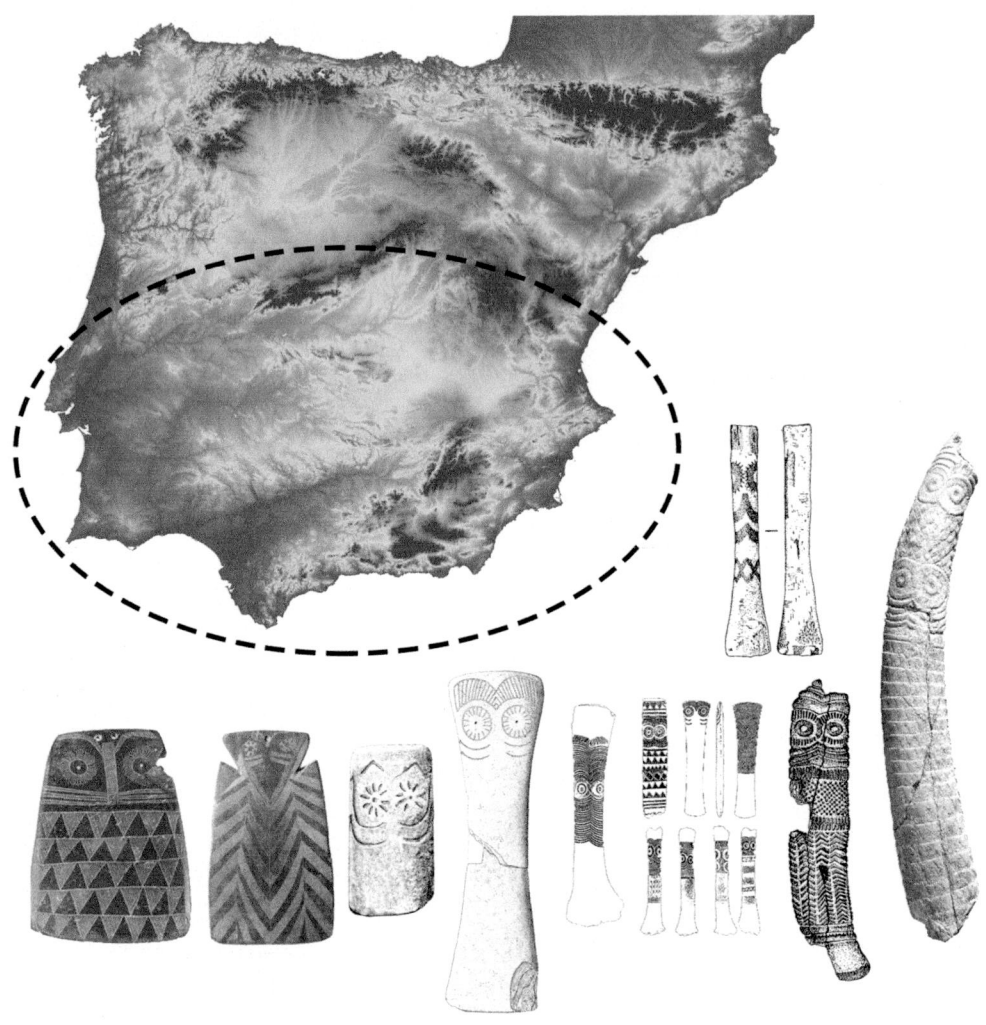

FIGURE 7. THE EYE-MOTIF IS ADOPTED AND ADAPTED TO A VARIETY OF SUPPORTS OVER A WIDE GEOGRAPHICAL FRAMEWORK, WITH REGIONAL DIFFERENCES. THE DISTRIBUTION OF THESE FIGURATIONS SHOWS THE FLOW OF A SHARED IDEA, HIGHLIGHTING CONNECTIONS BETWEEN REMOTE AREAS.

The distributions we have referred to are just some examples that provide evidence of the circulation not only of materials and objects of different types but also of an intangible element like information. These circuits are an expression of contacts and relationships of various types that involve groups both neighbouring and distant. However, tracing these distributions by reading the archaeological record is not always enough and a more detailed analysis needs to be carried out to enable us to understand the economic and social phenomena at the root of these relationships.

Exchange and Interaction. Contacts and Social Relationships

This circulation of products could involve:

- the exchange of manufactured goods or raw materials between various agents,
- the transfer of information and/or associated technology, or
- the movement of people.

All these possibilities speak to us of a single phenomenon: social interaction. It means the expression of dynamic relations between individuals and groups. The building of these links and their variations over time (continuing, extending or disappearing) could reveal changes or stability in societies. Like the intensification of production, exchanges are considered to play a part in the emergence of social inequality, and although it is true that all forms and models of exchange have an economic, political and/or social component, social complexity is not a prerequisite for these relationships.

The term *exchange* sometimes goes hand in hand with adjectives like 'prestigious' and 'exotic', referring to the character of the materials that circulate. A high social value is usually attributed to certain elements on the basis of their allochthonous origin, their scarcity or exoticism, or in consideration of the skills and knowledge involved in their manufacture. The value given to particular objects as elements of prestige is related to restricted ownership, increasing the status of any individuals or groups that have access or knowledge. However, it is difficult to imagine that the origin of these circuits used to distribute materials, knowledge and/or people could be linked to excess production and to controlling and maintaining inequalities. Some of the examples given in the previous section for the early stages of the Neolithic illustrate the importance of the social motivations (interpersonal relationships or connections, the establishment of alliances) behind these exchanges for the groups of humans that participate. This is the case with the circulation between the south-east of the Peninsula and the Valencia area of polished tools made of amphibolite, instruments with similar features to tools made of stone from the regional framework. The use of these axes is not considered to be restricted to one part of the group and neither is it associated with a particular context (Orozco Köhler, 2000). The social value of these objects speaks to us of the affective meaning of these ties or connections. However, during the earliest moments of the Neolithic, researchers have also documented the circulation between the same regions of elements of personal adornment, such as schist bracelets, whose scarcity in the record and limited presence in certain archaeological contexts (Bernabeu *et al.*, 2006) increases their singularity and high social value. In general terms we can assume that two elements of different types and different functions can be involved in the same circuit. However, it is possible that their distribution could respond to different needs and realities.

Moving forward through the time scale we will find other materials with presumably different functions and values which are involved in interrelated circuits Thus during the IV and III millennia cal BC, the central and southern Iberian Peninsula seems to delimit an area within which informa-tion, objects and possibly people circulate, forming circuits of different scales, natures and intensities.

On a general scale, the distribution of the eye-shaped motifs mentioned earlier (Hurtado, 2008) tells us of the existence of extensive interregional circuits via which information circulates. And with the passing of time, ivory (Schuhmacher *et al.*, 2009) and the earliest Bell Beaker pottery appear to use and even extend these same circuits, placing the region into a wider Mediterranean context (Bernabeu and Molina, 2011).

On a more limited scale, various studies point to the circulation of raw materials and/or finished products, such as the obsidian pieces that reach the Catalan area (Gibaja *et al.*, 2014), the variscite from Pico Centeno (Odriozola *et al.*, 2010; 2013) that circulates in areas of the south-west, and certain types of tough stone that are distributed throughout the eastern coastline of the Iberian Peninsula (Orozco, 2000, 2011). These are circuits to which we could also add the spread of metal and metallurgy, according to some recent results (Rovira and Montero, 2011).

All this clearly reflects a situation that is more common than is usually imagined: the superimposition of exchange networks. Over the same physical space we see a flow of different types of material (variscite, amphibolite, ivory, metal and others). Sometimes this flow carries something more subtle, like information relating to the technological (metallurgy) or ideological areas (the eye-shaped motif or the earliest Bell Beaker pottery, which follows and even extends this same circuit). The flow of information takes place simultaneously but affects different parts of the geography through which it passes. More importantly, it contains a dynamic component in that it changes over time, although, following the approaches usually used to analyse these flows, we have so far been unable to explain precisely how they work or how they evolve.

These are indeed processes of exchange, but behind this concept it can be guessed that there are interactions between different types of agent forming a network that changes over time (dynamic) and affects the space, and which is in turn affected by the way these agents occupy and organize themselves within the space. This exchange is more than just the distribution of objects whose origins we know (sometimes). It also includes the circulation of ideas and immaterial innovations whose origin is usually subject to more debate, but the materiality of which is reflected in various aspects of the material culture.

Understanding the dynamic means going beyond the concepts commonly used to analyse forms of exchange. Certain concepts and perspectives that are frequently used in the area of the social sciences are timidly being applied in Archaeology. Social Network Analysis (Knappet, 2011) and the study of Complex Systems are tools that are beginning to gain importance in prehistoric studies and are proving to be efficient in providing information about past societies (Bernabeu *et al.*, 2012). Their potential for studies into exchange is obvious, especially as regards the processes involved in the transfer of information. This would make it possible to understand the form, intensity and extension (scale) of some of these circuits, including their fragmentation (Mills *et al.*, 2013). A path is opening up that will almost certainly widen our knowledge of recent prehistory.

References

Bernabeu, J.; Moreno, A.; Barton, C. M., 2012 – Complex Systems, Social Networks and the evolution of Social Complexity. In: Cruz Berrocal, M.; García Sanjuan, L. and Gilman, A. (eds): The Prehistory of Iberia. Debating Early Social Stratification and the State. Routledge: pp. 23-37.

Bernabeu, J.; Molina, Ll., 2011 – El Horizonte Campaniforme 30 años después. In: Pérez, G.; Bernabeu, J.; Carrión, Y.; García, O.; Molina, Ll. and Gómez, M. (eds): La Vital (Gandía, Valencia). Vida y muerte en la desembocadura del Serpis durante el III y el I milenio a.C. Serie Trabajos Varios del SIP, 113. Diputación de Valencia: pp. 275-279.

Bernabeu, J.; Molina, Ll.; Diez, A.; Orozco, T., 2006 – Inequalities and Power. Three millenia of Prehistory in mediterranean Spain (5600-2000 calBC). In: Díaz-del-Río, P. and García Sanjuan, L. (eds): Social Inequality in Iberian Late Prehistory. British Archaeological Reports, IS. 1525: pp. 97-116.

Capote, M.; Castañeda, N.; Consuegra, S.; Criado, C.; Diaz-del-Rio, P., 2008 – Flint mining in early neolithic Iberia: a preliminary report on 'Casa Montero' (Madrid, Spain). In: Allard, P.; Bostyn, F.; Giligny, F.; Lech, J. (eds): Flint mining in Prehistoric Europe. Interpreting the archaeological records. British Archaeological Reports, IS. 1891: pp. 123-137.

Diaz-del-Rio, P.; Consuegra, S.; Capote, M.; Castañeda, N.; Criado, C.; Orozco, T.; Terradas, X., 2008 – Estructura, contexto y cronología de la mina de sílex de Casa Montero (Madrid). IV Congreso del Neolítico en la Península Ibérica, (Alicante, 2006), vol. 1. Museo Arqueológico de Alicante. Alicante: pp. 200-207.

Diaz-del-Rio, P.; Consuegra, S.; Capdevila, E.; Capote, M.; Casas, C.; Castañeda, N.; Criado, C.; Nieto, A., 2010 – The Casa Montero Flint mine and the Making of Neolithic Societies in Iberia. In: Anreiter, P.; Goldenberg, G.; Hanke, K.; Krause, R.; Leitner, W.; Mathis, F.; Nicolussi, K.; Oeggl, K.; Pernicka, E.; Prast, M.; Schibler, J.; Schneider, I.; Stadler, H.; Stöllner, T.; Tomedi, G.; Tropper, P. (eds): Mining in European History and its impact on Environment and Human Societies. Proceedings for the 1st Mining in European History-Conference of the SFB-HIMAT (Innsbruck 2009): pp. 351-355. Innsbruck University Press.

Diaz-del-Rio, P. and Consuegra, S., 2011 – Time for action. The chronology of mining events at Casa Montero (Madrid, Spain). In: Capote, M.; Consuegra, S.; Diaz-del-Rio, P. and Terradas, X. (eds)

Proceedings of the 2nd Internacional Conference of the UISPP Commission on Flint Mining in Pre- and Protohistoric Times (Madrid, 2009). British Archaeological Reports, IS. 2260: pp. 221-229.

Gibaja, J. F.; Ibañez, J. J.; Rodríguez, A.; González, J. E.; Clemente, I.; García, V.; Perales, U., 2010 – Estado de la cuestión de los estudios traceológicos realizados en contextos mesolíticos y neolíticos del sur peninsular y noroeste de África. Os últimos caçadores-recolectores e as primeiras comunidades produtoras do sul da Península Ibérica e do norte de Marrocos. Promontoria Monográfica, 15: pp. 181-189.

Gibaja, J. F.; González, P.; Martin, A.; Palomo, A.; Petit, M. A.; Plasencia, X.; Remolins, G.; Terradas, X., 2014 – New finds of obsidian bladelets at Neolithic sites in north-eastern Iberia. Antiquity Project Gallery, p. 340.

Goñi, A.; Rodríguez, A.; Camalich, M. D.; Martín, D.; Francisco, M. I., 1999 – La tecnología de los elementos de adorno personal en materias minerales durante el Neolítico Medio. El ejemplo del poblado de Cabecicos Negros (Almería). II Congrés del Neolític a la península Ibérica. Sagvntvm-Extra II: pp. 163-170.

Goñi, A.; Chávez, E.; Camalich, M. D.; Martín, D.; González, P., 2000 – Intervención arqueológica de urgencia en el poblado de Cabecicos Negros (Vera, Almería). Informe preliminar. Anuario Arqueológico de Andalucía 2000. III-1: pp. 73-87. Junta d Andalucía.

Hurtado, V., 2008 – Ídolos, estilos y territorios de los primeros campesinos en el sur peninsular. In: Cacho, C.; Maicas, R.; Martos, J. A. and Martínez, M. I. (coord) Acercándonos al pasado. Prehistoria en 4 actos. Ministerio de Cultura. MAN and CSIC. http://man.mcu.es/museo/JornadasSeminarios/acercandonos_al_pasado.html

Knappett, K., 2011 – An Archaeology of Interaction. New Perspectives on Material Culture and Society. Oxford University Press.

Mills, B.; Clark, J. J.; Peeples, M. A.; Haas, W. R.; Roberts, J.; Hill, J.; Huntley, D. L.; Borck, L.; Breiger, R. L.; Clauset, A.; Shackley, M. S., 2013 – Transformation of social networks in the late pre-Hispanic US Southwest. PNAS, 110, 15: pp. 5785-5790.

Linares, J. A. and Odriozola, C., 2011 – Cuentas de collar de variscita y otras piedras verdes en tumbas megalíticas del suroeste de la Península Ibérica. Cuestiones acerca de su producción, circulación y presencia en contextos funerarios. Menga, 1: pp. 335-369.

Odriozola, C. P.; Linares-Catela, J. A.; Hurtado-Pérez, V., 2010 – Variscite source and source analysis: testing assumptions at Pico Centeno (Encinasola, Spain). Journal of Archaeological Science, 37: pp. 3146-3157.

Odriozola, C.; Sousa, A. C.; Boaventura, R.; Villalobos, R., 2013 – Componentes de adornos de pedra verde de Vila Nova de São Pedro (Azambuja): Estudo de proveniências e redes de troca no 3º milenio A.N.E. no actual território portugués. Arqueologia em Portugal. 150 anos: pp. 457-462.

Orozco Köhler, T., 2000 – Aprovisionamiento e Intercambio. Análisis petrológico del utillaje pulimentado en la Prehistoria reciente del País Valenciano (España). British Archaeological Reports, IS. 867. Oxford.

Orozco Köhler, T., 2011 – Materiales líticos no tallados. In: Pérez, G.; Bernabeu, J.; Carrión, Y.; García, O.; Molina, Ll. y Gómez, M. (eds): La Vital (Gandía, Valencia). Vida y muerte en la desembocadura del Serpis durante el III y el I milenio a.C. Serie Trabajos Varios del SIP, 113. Diputación de Valencia: pp. 175-181.

Pétrequin, P.; Cassens, S.; Errera, M.; Klassen, L.; Sheridan, A., 2012 – JADE. Grandes haches alpines du Néolithique européen. V et IV millénaires av. J-C. 2 vol. Presses Universitaires de Franche-Comté. Beçanson.

Rovira, S. and Montero-Ruiz, I., 2011 – Aspectos metalúrgicos. In: Pérez, G.; Bernabeu, Carrión, J. Y.; García, O.; Molina, Ll. y Gómez, M. (eds): La Vital (Gandía, Valencia). Vida y muerte en la desembocadura del Serpis durante el III y el I milenio a.C. Serie Trabajos Varios del SIP, 113. Diputación de Valencia: pp. 219-227.

Sharckley, M. S., 2008 – Archaeological petrology and the archaeometry of lithic materials. Archaeometry, 50 (2): pp. 194-215.

Schuhmacher, T. X.; Cardoso, J. L.; Banerjee, A., 2009 – Sourcing African ivory in Chalcolithic Portugal. Antiquity, 83: pp. 983-997.

Terradas, X.; Gratuze, B.; Bosch, J.; Enrich, R.; Esteve, X.; Oms, F. X.; Ribé, G., 2014 – Neolithic diffusion of obsidian in the western Mediterranean: new data from Iberia. Journal of Archaeological Science, 41: pp. 69-78.

The Western network revisited: the transition into agro-pastoralism in the Alto Ribatejo, Portugal

Nelson J. Almeida[1,2,3], Cristiana Ferreira[1,2,3], Sara Garcês[1,2,3], Ana Cruz[4,2,3], Pierluigi Rosina[5,2,3], Luiz Oosterbeek[5,2,3]

[1] University of Trás-os-Montes and Alto Douro (UTAD, Portugal)
[2] Quaternary and Prehistory Group, Geosciences Centre, University of Coimbra (GQP-CG, UC, ulandD73, FCT, Portugal); Institute for Interdisciplinary Research, University of Coimbra (IIIUC, Portugal)
[3] Earth and Memory Institute (ITM, Portugal)
[4] Prehistory Centre, Polytechnic Institute of Tomar (CPH, IPT, Portugal)
[5] Polytechnic Institute of Tomar (IPT, Portugal)

Abstract

The Alto Ribatejo (central Portugal) is a sub-region with an important amount of data for the study of the neolithisation process. Some aspects related to contexts and chronologies, vegetation dynamics, palaeoeconomy and art records are reviewed. Published interpretations and models are discussed taking account of existing data, allowing pointing out some research problems and future directions.

Key words: *Neolithisation, Tagus Valley, Alto Ribatejo*

Résumé

Le Alto Ribatejo (centre du Portugal) est une sous-région avec une quantité importante de données pour l'étude du processus de néolithisation. Certains aspects liés aux contextes et chronologies, dynamiques de la végétation, paléoéconomie et registres d'art sont examinés. Interprétations et modèles publiés sont discutés en tenant compte des données existantes, permettant d'indiquer certains problèmes et orientations futures de recherche.

Mots clés: *Néolithisation, Vallée du Tage, Alto Ribatejo*

1. Introduction

The neolithisation of Western Iberia has been a matter of discussion for several decades. From earlier interpretations to more recent ones (Guilaine and Veiga Ferreira, 1970; Arnaud, 1982; Zilhão, 2001; Carvalho, 2008), most scholars seem to agree to the arrival of allochtonous pioneer groups bearing certain elements of the 'Neolithic way of life'. Accepting these initial moments of neolithisation of the coastal areas of Portugal, other aspects related to the inner areas need to be addressed, namely questions related to the occupation or inoccupation of these areas by epipalaeolithic/mesolithic hunter-gathers and, in the case of occupation, of how did these different groups interacted.

The Alto Ribatejo is an interesting area in which to discuss the above-mentioned questions: it may comprise exogenous dynamics in the beginning of the neolithisation (Zilhão, 2001; Carvalho, 2008) and various degrees of endogenous influences (Oosterbeek, 1994; Cruz, 1997, 2011, in press) regarding the development of the regional Neolithic.

2. Geological and geographical setting

Alto Ribatejo is a sub-region of Central Portugal (western Iberian Peninsula) with an area of about ~2.500 km², located along the banks of the lower Tagus (fig. 1). The hydrographic system is controlled by tectonics, so the Tagus River dominates the hydrographical system in this area (east-west), followed by the Zêzere River, which flows into the Tagus from the north and east, and the Nabão River, which flows into the Zêzere from the north.

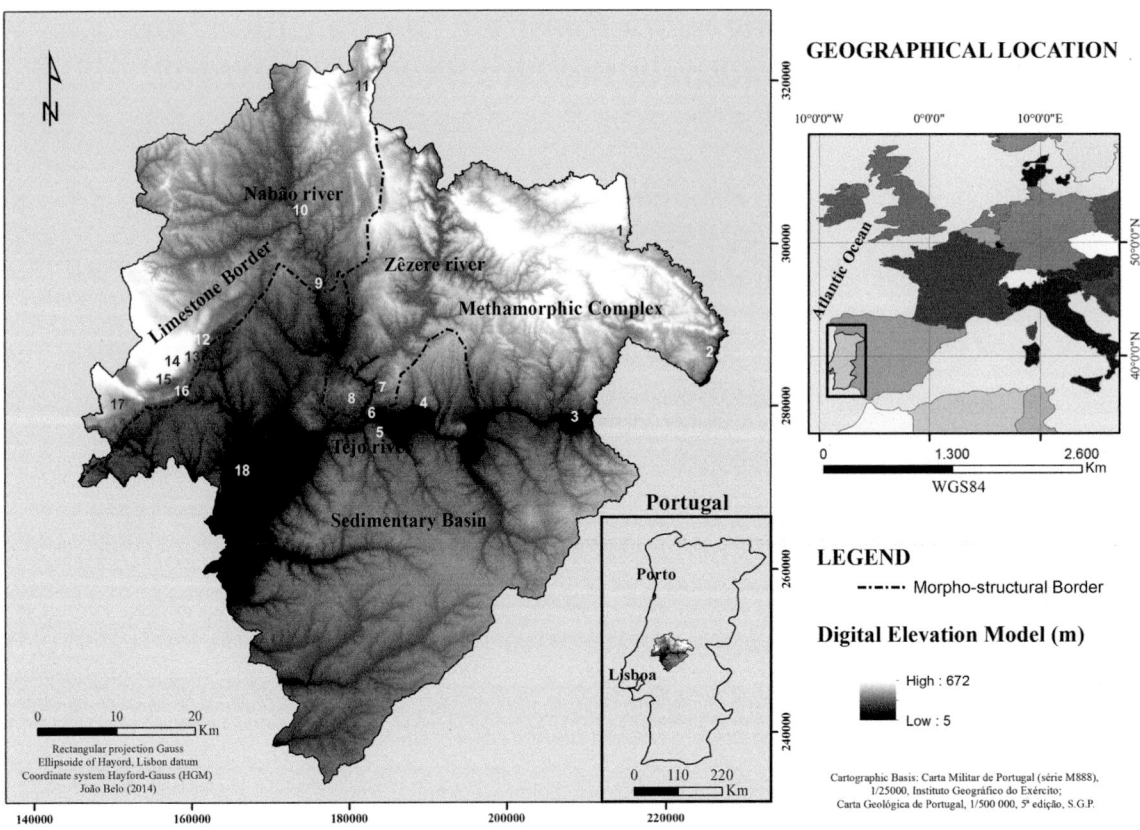

FIGURE 1. DIGITAL ELEVATION MODEL OF THE ALTO RIBATEJO WITH LOCATION OF THE SITES MENTIONED IN THE TEXT. 1. ANTA DA LAJINHA; 2. OCREZA ROCK ART COMPLEX; 3. ANTA DO RIO FRIO; 4. AMOREIRA; 5. SANTA MARGARIDA DA COUTADA; 6. PEDRA DA ENCAVALADA; 7. FONTES; 8. ANTA 1 DE VAL DA LAJE; 9. FONTE QUENTE; 10. CALDEIRÃO, CADAVAL, OSSOS, Nª Sª LAPAS; 11. REGO DA MURTA; 12. PENA D'ÁGUA; 13. COSTA DO PEREIRO; 14. PICAREIRO; 15. CERRADINHO DO GINETE; 16. CISTERNA; 17. CARRASCOS, GALINHA; 18. PAÚL DO BOQUILOBO.

The geomorphology is quite diversified because of the different bedrock. In fact, in this region there are three main geological units: the Cenozoic Tagus sedimentary basin (in the geological map 'Tagus and Sado Tertiary Basin'), that extends itself mostly along the lower Tagus river valley of central Portugal; the sedimentary basin in the Northwest and West is limited by the 'Estremenho' Limestone Massif, essentially Mesozoic; and in the Northeast and the East, by the Pre-Cambrian and Palaeozoic schist-metamorphic complex (Ancient Massif).

Lithologically, there are limestones and marls (with weak presence of flint) in the 'Estremenho' Massif. Schist, greywacke, quartzite and granite are the more abundant rocks in the Ancient Massif. Finally clay, silts, sands, and pebbles make up the detritic drainage basin. The Holocene alluvial sediments, the Pleistocene wide fluvial terraces, the karstic cave fillings (Limestone Massif), and the detritic coverings represent the regional Quaternary deposits. The Holocene alluvial sediments arrive at a depth of more than 9 meters in the Tagus sedimentary basin.

3. Evidences and interpretations

3.1. Contexts and chronologies

Besides several surface scatters previously identified, the regional Holocene hunter-gatherers have a low representativeness: datings obtained in the eastern areas of the Alto Ribatejo, in the open-air settlements of Fontes and Amoreira, are indicative of Pre-Boreal and/or Boreal chronologies (Oosterbeek, 1994; Cruz, 1997, 2011) (table 1). Other sites located in the Limestone Massif also present occupations for these periods: Picareiro (Bicho *et al.*, 2003), Pena d'Água and Costa do Pereiro (Carvalho, 1998, 2008). Post-glacial macrolithic industries are seen in some of these sites, with a higher predominance in the eastern ones but quartzite being also common in the western areas.

Regional Early Neolithic records comprise paradigmatic sites with cardial and epicardial ceramics located in the limestone area, as Caldeirão (Zilhão, 1992), Cisterna (Zilhão, 2001), Pena d'Água (Carvalho, 1998) and Nª Sª das Lapas (Oosterbeek, 1994). Further east, besides cardial ware identified in surface scatters (Silva *et al.*, 2009), the settlement of Fontes also presents cardial decorated vessels (Cruz, 2011). Still, the TL datings obtained for the sediment (layer C) suggest a 'contemporaneity' between plain and impressed decorated ceramics in the region (Cruz, 2013). Regarding lithic raw materials, the few Early Neolithic sites located in the eastern areas present a clear predominance of quartzite that also occurs in the western areas, in the Limestone Massif, but in lower percentages.

Regional Early/Middle Neolithic sites relate mainly to funerary episodes and/or punctual specific occupations, lacking larger occupations (besides Fontes or Amoreira). The Middle Neolithic is represented by several contexts, with a majority of them located in the Limestone Massif, but also an important amount of megalithic contexts in the remaining areas. The latter are less precise in terms of their chronology, being possible to also represent Final Neolithic chronologies. For the Middle/Late Neolithic, some sites should be mentioned, as Cadaval (Oosterbeek, 1994) and Carrascos (Gonçalves

Site	Lab. Ref.	Sample	BC	BP	cal BC 2s
Fontes*	ITN-LUM-452	Clay structure	8600±600		
Fontes*	ITN-LUM-451	Clay structure	8100±600		
Fontes*	ITN-LUM-453	Clay structure	8000±600		
Amoreira	Beta-189993	Charcoal		9010±40	8300-7990
Picareiro	Wk-6676	Charcoal		8310±130	7580-7060
Pena d'Água	Wk-9213	*Quercus suber*		7370±110	6440-6030
Costa do Pereiro	Wk-17026	*Cervus elaphus*		7327±42	6340-6060
Fontes*	ITN-LUM-450	Colluvium/Soil	5700±500		
Almonda	OxA-9287	Adornment *Cervus*		6445±45	5490-5320
Almonda	OxA-9288	Adornment Bone		6445±45	5490-5320
Caldeirão	OxA-1035	*Ovis aries*		6390±80	5480-5070
Caldeirão	OxA-1034	*Ovis aries*		6230±80	5370-4980
Caldeirão	OxA-1033	*Homo*		6130±90	5300-4840
Caldeirão	OxA-1037	*Bos taurus*		5970±120	5220-4580
Nª Sª Lapas	ICEN-802	*Homo*		6100±70	5220-4840
Caldeirão	OxA-1036	*Bos taurus*		5870±80	4940-4540
Caldeirão	TO-350	*Homo*		5810±70	4830-4490
Pena d'Água	Wk-16418	*Olea europaea*		5831±40	4800-4550
Cadaval	ICEN-803	*Homo*		5390±50	4350-4050
Cadaval	I-17241	*Homo*		5180±140	4330-3700
Nª Sª Lapas	I-17247	*Homo*		5130±140	4310-3650
Cadaval	ICEN-464	*Homo*		5160±50	4150-3790
Costa do Pereiro	Wk-13682	*Homo*		5133±45	4040-3790
Cadaval	Beta-189995	*Homo*		4640±40	3620-3350

TABLE 1. DATINGS FOR THE MAIN CONTEXTS REFERRED IN THE TEXT, ACCORDING TO OOSTERBEEK, 1994; CRUZ, 1997, 2011; BICHO *ET AL.*, 2003; ZILHÃO, 2001; CARVALHO, 2008 AND BURBIDGE *ET AL.*, 2014. CONVENTIONAL AND AMS C14 DATINGS WERE CALIBRATED WITH INTCAL13 CALIBRATION CURVE IN THE OXCAL (VERSION 4.2.3) PROGRAM (BRONK RAMSEY, 2009; REIMER *ET AL.*, 2013). * TL DATINGS.

and Pereira, 1974/1977) in the Limestone Massif, and several megalithic monuments (e.g., Anta 1 de Val da Laje, Anta da Lajinha, Anta do Rio Frio) in the eastern areas (Scarre and Oosterbeek, 2010).

Throughout the Neolithic, the type of locally available raw materials and exchanges networks conditioned the typological diversity of lithic artefacts: quartzite and quartz are the dominant raw materials, abundantly found in the tertiary and quaternary gravels of the Tagus and Zêzere river valleys, and relatively frequent in the Nabão river valley. Quartzite and quartz petrographic characteristics and the form that they are present (e.g., small to medium pebbles, large clasts, nodules) affected the application of several knapping techniques and reduction sequences. Overall, the frequency of products over flakes is higher than the evidences of unifacial and bifacial pebbles, whose exploration was oriented according to the *débitage* axis (in the majority of the cases coincident with the longer morphological axis), allowing the configuration of unidirectional and bidirectional knapped pebbles, as well as nucleus. Regional Holocene industries are indicative of a continuity or at least convergence, mainly in the eastern areas (Grimaldi *et al.*, 1999; Cruz *et al.*, 2000). Amphibolite, a local raw material in the valleys of the Tagus and Zêzere rivers and non-local in the Nabão river valley, was recurrently used to produce polished axes, gouges and adzes. Although sporadically found in the form of small nodules in the Tagus fluvial deposits, flint may be considered a non-local raw-material this way allowing to talk of use and provision strategies. Its exploration aimed to obtain laminar and lamellar products, following one or several *chaîne opératoire*, conceptually oriented towards prismatic geometries.

As for ceramic vessels shapes, they were produced through the coiling technique, modelling and moulding. The surfaces treatments is variable with slips engobes, smoothing and polishing being present, followed by incise, impressed, punctured, combed (forming geometrical motifs similar to the ones found on other areas of Portugal), burnished or plastic decoration being represented. Primary shapes and some composite shapes, easily chronologically identifiable to the Early Neolithic are found. During the Middle-Final Neolithic bowl shaped, spherical, ovoid and conic shapes occur. Two main clusters seem to exist in the regional Neolithic: on the western areas (e.g., Caldeirão, Nª Sª das Lapas, Cadaval), related to the *circum*-Mediterranean influences, consisting mainly on relatively large, hard and decorated vessels, representing the initial appearance of ceramic in the region; posteriorly, the eastern cluster (e.g., Fontes, Amoreira, Anta 1 de Val da Laje) with mainly small brittle and plain ceramics with no cooking possible purposes and very limited storage capacity. This second cluster offers stronger links with inland occupations, both the megalithic network and the prior mobile late hunters and shepherds.

3.2. Palaeobotany

A gradual alteration of the landscape throughout the Holocene is observed in the decline of woodland species (*Quercus* and *Pinus*) and the increase of shrubs and herbaceous. Although primarily associated to an increasing anthropogenic impact during the Middle Holocene (Allué, 2000; Ferreira, 2010) thus following the dominant view in western Iberia (Mateus and Queiroz, 1993; Zilhão 1992), a possible climate influence on the deforestation was not discarded, as reported in other regions (Burjachs, 1990; Jalut *et al.*, 1997; Burjachs and Allué, 2003; Grau and Duque, 2007).

During the Epipalaeolithic/Early Neolithic, a major representativeness of woodland species (*Pinus*, deciduous *Quercus*, *Alnus* and *Olea europaea*), and low values of shrubs (Ericaceae, Cistaceae) and herbaceous (wild grass) is evident (table 2) (Figueiral, 1998; Allué, 2000; Bicho *et al.*, 2003; Vis *et al.*, 2010; Ferreira *et al.*, 2014). Throughout the Middle Neolithic a reduction of deciduous *Quercus* and *Pinus* occurs; at the same time, shrub species began to establish themselves more intensively in the landscape (Vis *et al.*, 2010). *Olea europaea* is the most representative taxon, accompanied by sclerophillous species (*Quercus* evergreen, Ericaceae). Pollen studies show considerable values of Asteraceae (open areas) and an increasing representation of shrubs and other herbaceous (Poaceae) possibly correlated with activities of agricultural and farm production. Other authors suggest that the

Taxon	Epipalaeolithic			Early Neolithic	Middle Neolithic				Late Neolithic	Chalcolithic		
	APA*	Amr*	PBoq^	APA*	PBoq^	VL^*	APA*	GSL*	APA*	PBoq	VL^*	SMAC^
Pinus sp.		+++	++		++	+				+	+	
Deciduous *Quercus*			++		++	+	+		+	+		+
Evergreen *Quercus*	+	+	++	+	++	+	+	++	+	++	+	+
Olea europaea	+++			+++		+	+++	+	++	+	+	
Riparian woodland			+++		++				+	++	+	
Ericaceae						+				+++		+
Arbutus unedo		+				++		+	+		+	
Erica sp.			+		++	+	+		+	++	++	
Cistaceae	+	+	+		+	+				+	+	
Rhamnus alaternus/*Phillyrea*	+			+		+		+	+		+	
Vitis		+			+							
Poaceae		++			+++					+++	+	++
Asteraceae		++			++	++				+	+	++
Plantago sp.						+				+		+
Cerealia – type						+						+++

TABLE 2. GENERAL ABUNDANCE OF THE MAIN TAXA IDENTIFIED IN THE ALTO RIBATEJO PALYNOLOGICAL(^) AND ANTHRACOLOGICAL(*) RECORDS (+ 0-25%; ++ 25-50%; +++ >50%). APA = PENA D'ÁGUA (FIGUEIRAL, 1998); AMR = AMOREIRA (ALLUÉ 2000; FERREIRA ET AL., 2014); PBoQ = PAÚL DO BOQUILOBO (VIS ET AL., 2010); VL = ANTA 1 DE VAL DA LAJE (ALLUÉ, 2000); GSL = Nª Sª DAS LAPAS (ALLUÉ, 2000); SMAC = SANTA MARGARIDA DA COUTADA (FERREIRA, 2010).

first anthropic markers may go back to 6000 BP and, even though the development of scrubs of Erica dates back to 5000 BP (Mateus and Queiroz, 1993), species of mesophyll character are increasingly encountered since this phase.

In the Final Neolithic/Chalcolithic we can see the maximum values for the populations of Ericaceae. *Pinus* has a very low representativeness and deciduous *Quercus* presents a gradual decrease. *Vitis*, *Olea*, wild grasses and other grasses associated to farmlands present a higher representativeness; also, cereals appear in the palynological studies. Ericaceae (*Arbutus unedo*, *Erica arborea*, *Calluna*) remain the most representative taxa, but other shrubs (Cistaceae, *Pistacia*, Leguminosae, *Ligustrum*, and *Myrtus*) were identified (Figueiral, 1998; Allué, 2000).

3.3. Zooarchaeology

Epipalaeolithic/Mesolithic archaeofauna is registered in regions west of Alto Ribatejo (Detry, 2007, Valente, 2008). Regional Neolithic zooarchaeological data is quite sparse and circumscribed to the Limestone Massif (Almeida *et al.*, 2014; Valente and Carvalho, 2014). Early Neolithic data is to be found in Caldeirão, Pena d'Água and Cerradinho do Ginete (table 3). In the first, *Sus scrofa* predominates in terms of NISP values (Rowley-Conwy, 1992; cf. Davis, 2002; Albarella *et al.*, 2005), but *Bos taurus* and *Ovis aries* (as well as *Ovis aries*/*Capra hircus*) are also present. Pena d'Água presents a predominance of *Cervus elaphus*, followed by *Ovis aries*/*Capra hircus*, *Bos* sp. and *Sus* sp.

Middle Neolithic data is to be found in Pena d'Água, Costa do Pereiro and Cadaval where a predominance of *Cervus elaphus* or *Ovis aries*/*Capra hircus* is seen; other taxa (i.e., *Bos* sp., *Sus* sp.) have vestigial values. As for the Final Neolithic, regional data concerns only Cadaval and Morgado superior, but chronological revisions are on-going. While *Sus* sp. remains difficult to specifically

Taxon	Early Neolithic	Middle Neolithic	Final Neolithic
Bos sp.	++	+	+
Cervus elaphus	++	+++	+
Sus sp.	+++	++	+
Capreolus capreolus	+	+	+
Ovis aries/Capra hircus	++	+++	++

TABLE 3. GENERAL ABUNDANCE OF THE MAIN TAXA IDENTIFIED IN THE ALTO RIBATEJO ZOOARCHAEOLOGICAL RECORDS BASED ON NISP VALUES (+ <10; ++ 10-50; +++ >50). EARLY NEOLITHIC: CALDEIRÃO (ROWLEY-CONWY, 1992), PENA D'ÁGUA (VALENTE, 1998; CARVALHO ET AL., 2004; CARVALHO, 2008), CERRADINHO DO GINETE (CARVALHO ET AL., 2004); MIDDLE NEOLITHIC: PENA D'ÁGUA (VALENTE, 1998; CARVALHO, 2008; LUÍS ET AL., 2013), COSTA DO PEREIRO (CARVALHO, 2008), CADAVAL (ALMEIDA ET AL., IN PRESS). FINAL NEOLITHIC: CADAVAL AND MORGADO SUPERIOR (ALMEIDA ET AL., IN PRESS).

identify (e.g., Albarella *et al.*, 2005), *Bos* sp. corresponds to the domestic variant whenever a specific adscription is possible, exception being one tooth identified in the Early Neolithic of Pena d'Água, with some reservations (Carvalho, 2008; Valente and Carvalho, 2014).

3.4. Rock art and other manifestations

In the western areas of the Alto Ribatejo, symbolic evidences correspond mainly to portable features, as schist geometrically engraved plaquettes and shell or stone ornaments (bracelets, beads, pendants). An example is the Carrascos cave, where personal adornment objects represent a high percentage of the material found; or the Galinha cave, where interesting schist plaquettes were found, along with other Neolithic materials (Gonçalves and Pereira, 1974/1977). In contrast, no evident purely 'symbolic' items are found in the eastern areas, even in the first megalithic burials (Oosterbeek *et al.*, 2014). Here, symbolic representations are concentrated in the Tagus Valley Rock Art Complex, where several rock art sites (Serrão, 1972) occur substantially from the Cedilho dam in Spain to the mouth of the Ocreza River (Garcês, 2009), reaching some tributary rivers. The typology of engravings is very diverse: zoomorphic, anthropomorphic and a majority of geometric figures (e.g., circles, concentric circles, lines, cup-marks).

During the last few years there is a tendency to characterize some motifs that come across the Iberian Peninsula that do not exactly fit into the naturalism of Palaeolithic rock art neither on the Neolithic schematic features (Collado-Giraldo, 2004, 2006; Bueno-Ramírez *et al.*, 2009). Collado proposes an integration of these figures on a new rock art artistic cycle denominated *pre-schematic rock art* (Collado-Giraldo, 2004, 2006). This *style* breaks with the traits of Magdalenian filiform features and is characterized by the almost exclusive use of pecking but it also emerges figurative painted elements (Collado-Giraldo and García-Arranz, 2006).

The Tagus valley presents a wide chronology where evidences increasingly show us a sequence between Palaeolithic figures (Baptista, 2001), the figures of the Zêzere valley and post-Palaeolithic figures. These pre-schematic features could be identified due to figures like the deer in rock number 155 from Fratel. The motifs present in this rock are mainly big figures of animals with a diverse interior body decoration (x-rays and lines). All male deer antlers are found in perspective and, in spite of the body being displayed in profile, in six of these animal representations the front and back legs are also in perspective. Different ways of animal representation are patent: large animals profusely engraved and decorated internally with reticulated lines, wide necks and well-developed antlers or manes, and cervical-dorsal lines well pronounced, representing a naturalism that follows previous periods (M. V. Gomes sub-naturalism period); animals represented so much simpler than the previous, less obvious body lines (emerge with a more oval shape), interior decoration summarized only by

Taxon	Epipalaeolithic			Early Neolithic	Middle Neolithic				Late Neolithic	Chalcolithic		
	APA*	Amr*	PBoq^	APA*	PBoq^	VL^*	APA*	GSL*	APA*	PBoq	VL^*	SMAC^
Pinus sp.		+++	++		++	+				+	+	
Deciduous *Quercus*			++		++	+	+		+	+		+
Evergreen *Quercus*	+	+	++	+	++	+	+	++	+	++	+	+
Olea europaea	+++			+++		+	+++	+	++	+	+	
Riparian woodland			+++		++				+	++	+	
Ericaceae						+				+++	+	
Arbutus unedo		+				++		+	+		+	
Erica sp.			+		++	+	+		+	++	++	
Cistaceae	+	+	+		+	+				+	+	
Rhamnus alaternus/ Phillyrea	+			+		+		+	+		+	
Vitis				+	+							
Poaceae			++		+++					+++	+	++
Asteraceae			++		++	++				+	+	++
Plantago sp.					+					+	+	
Cerealia – type					+							+++

TABLE 2. GENERAL ABUNDANCE OF THE MAIN TAXA IDENTIFIED IN THE ALTO RIBATEJO PALYNOLOGICAL(^) AND ANTHRACOLOGICAL(*) RECORDS (+ 0-25%; ++ 25-50%; +++ >50%). APA = PENA D'ÁGUA (FIGUEIRAL, 1998); AMR = AMOREIRA (ALLUÉ 2000; FERREIRA ET AL., 2014); PBOQ = PAÚL DO BOQUILOBO (VIS ET AL., 2010); VL = ANTA 1 DE VAL DA LAJE (ALLUÉ, 2000); GSL = Nª Sª DAS LAPAS (ALLUÉ, 2000); SMAC = SANTA MARGARIDA DA COUTADA (FERREIRA, 2010).

first anthropic markers may go back to 6000 BP and, even though the development of scrubs of Erica dates back to 5000 BP (Mateus and Queiroz, 1993), species of mesophyll character are increasingly encountered since this phase.

In the Final Neolithic/Chalcolithic we can see the maximum values for the populations of Ericaceae. *Pinus* has a very low representativeness and deciduous *Quercus* presents a gradual decrease. *Vitis*, *Olea*, wild grasses and other grasses associated to farmlands present a higher representativeness; also, cereals appear in the palynological studies. Ericaceae (*Arbutus unedo*, *Erica arborea*, *Calluna*) remain the most representative taxa, but other shrubs (Cistaceae, *Pistacia*, Leguminosae, *Ligustrum*, and *Myrtus*) were identified (Figueiral, 1998; Allué, 2000).

3.3. Zooarchaeology

Epipalaeolithic/Mesolithic archaeofauna is registered in regions west of Alto Ribatejo (Detry, 2007, Valente, 2008). Regional Neolithic zooarchaeological data is quite sparse and circumscribed to the Limestone Massif (Almeida *et al.*, 2014; Valente and Carvalho, 2014). Early Neolithic data is to be found in Caldeirão, Pena d'Água and Cerradinho do Ginete (table 3). In the first, *Sus scrofa* predominates in terms of NISP values (Rowley-Conwy, 1992; cf. Davis, 2002; Albarella *et al.*, 2005), but *Bos taurus* and *Ovis aries* (as well as *Ovis aries/Capra hircus*) are also present. Pena d'Água presents a predominance of *Cervus elaphus*, followed by *Ovis aries/Capra hircus*, *Bos* sp. and *Sus* sp.

Middle Neolithic data is to be found in Pena d'Água, Costa do Pereiro and Cadaval where a predominance of *Cervus elaphus* or *Ovis aries/Capra hircus* is seen; other taxa (i.e., *Bos* sp., *Sus* sp.) have vestigial values. As for the Final Neolithic, regional data concerns only Cadaval and Morgado superior, but chronological revisions are on-going. While *Sus* sp. remains difficult to specifically

Taxon	Early Neolithic	Middle Neolithic	Final Neolithic
Bos sp.	++	+	+
Cervus elaphus	++	+++	+
Sus sp.	+++	++	+
Capreolus capreolus	+	+	+
Ovis aries/Capra hircus	++	+++	++

TABLE 3. GENERAL ABUNDANCE OF THE MAIN TAXA IDENTIFIED IN THE ALTO RIBATEJO ZOOARCHAEOLOGICAL RECORDS BASED ON NISP VALUES (+ <10; ++ 10-50; +++ >50). EARLY NEOLITHIC: CALDEIRÃO (ROWLEY-CONWY, 1992), PENA D'ÁGUA (VALENTE, 1998; CARVALHO ET AL., 2004; CARVALHO, 2008), CERRADINHO DO GINETE (CARVALHO ET AL., 2004); MIDDLE NEOLITHIC: PENA D'ÁGUA (VALENTE, 1998; CARVALHO, 2008; LUÍS ET AL., 2013), COSTA DO PEREIRO (CARVALHO, 2008), CADAVAL (ALMEIDA ET AL., IN PRESS). FINAL NEOLITHIC: CADAVAL AND MORGADO SUPERIOR (ALMEIDA ET AL., IN PRESS).

identify (e.g., Albarella *et al.*, 2005), *Bos* sp. corresponds to the domestic variant whenever a specific adscription is possible, exception being one tooth identified in the Early Neolithic of Pena d'Água, with some reservations (Carvalho, 2008; Valente and Carvalho, 2014).

3.4. Rock art and other manifestations

In the western areas of the Alto Ribatejo, symbolic evidences correspond mainly to portable features, as schist geometrically engraved plaquettes and shell or stone ornaments (bracelets, beads, pendants). An example is the Carrascos cave, where personal adornment objects represent a high percentage of the material found; or the Galinha cave, where interesting schist plaquettes were found, along with other Neolithic materials (Gonçalves and Pereira, 1974/1977). In contrast, no evident purely 'symbolic' items are found in the eastern areas, even in the first megalithic burials (Oosterbeek *et al.*, 2014). Here, symbolic representations are concentrated in the Tagus Valley Rock Art Complex, where several rock art sites (Serrão, 1972) occur substantially from the Cedilho dam in Spain to the mouth of the Ocreza River (Garcês, 2009), reaching some tributary rivers. The typology of engravings is very diverse: zoomorphic, anthropomorphic and a majority of geometric figures (e.g., circles, concentric circles, lines, cup-marks).

During the last few years there is a tendency to characterize some motifs that come across the Iberian Peninsula that do not exactly fit into the naturalism of Palaeolithic rock art neither on the Neolithic schematic features (Collado-Giraldo, 2004, 2006; Bueno-Ramírez *et al.*, 2009). Collado proposes an integration of these figures on a new rock art artistic cycle denominated *pre-schematic rock art* (Collado-Giraldo, 2004, 2006). This *style* breaks with the traits of Magdalenian filiform features and is characterized by the almost exclusive use of pecking but it also emerges figurative painted elements (Collado-Giraldo and García-Arranz, 2006).

The Tagus valley presents a wide chronology where evidences increasingly show us a sequence between Palaeolithic figures (Baptista, 2001), the figures of the Zêzere valley and post-Palaeolithic figures. These pre-schematic features could be identified due to figures like the deer in rock number 155 from Fratel. The motifs present in this rock are mainly big figures of animals with a diverse interior body decoration (x-rays and lines). All male deer antlers are found in perspective and, in spite of the body being displayed in profile, in six of these animal representations the front and back legs are also in perspective. Different ways of animal representation are patent: large animals profusely engraved and decorated internally with reticulated lines, wide necks and well-developed antlers or manes, and cervical-dorsal lines well pronounced, representing a naturalism that follows previous periods (M. V. Gomes sub-naturalism period); animals represented so much simpler than the previous, less obvious body lines (emerge with a more oval shape), interior decoration summarized only by

Taxon	Epipalaeolithic			Early Neolithic	Middle Neolithic				Late Neolithic	Chalcolithic		
	APA*	Amr*	PBoq^	APA*	PBoq^	VL^*	APA*	GSL*	APA*	PBoq	VL^*	SMAC^
Pinus sp.		+++	++		++	+				+	+	
Deciduous *Quercus*			++		++	+	+		+	+		+
Evergreen *Quercus*	+	+	++	+	++	+	+	++	+	++	+	+
Olea europaea	+++			+++		+	+++	+	++	+	+	
Riparian woodland			+++		++				+	++	+	
Ericaceae						+				+++		+
Arbutus unedo		+				++		+	+		+	
Erica sp.			+		++	+	+		+	++	++	
Cistaceae	+	+	+		+	+				+	+	
Rhamnus alaternus/ Phillyrea	+			+		+		+	+		+	
Vitis			+		+							
Poaceae			++		+++					+++	+	++
Asteraceae			++		++	++				+	+	++
Plantago sp.					+					+		+
Cerealia – type					+							+++

TABLE 2. GENERAL ABUNDANCE OF THE MAIN TAXA IDENTIFIED IN THE ALTO RIBATEJO PALYNOLOGICAL(^) AND ANTHRACOLOGICAL(*) RECORDS (+ 0-25%; ++ 25-50%; +++ >50%). APA = PENA D'ÁGUA (FIGUEIRAL, 1998); AMR = AMOREIRA (ALLUÉ 2000; FERREIRA ET AL., 2014); PBOQ = PAÚL DO BOQUILOBO (VIS ET AL., 2010); VL = ANTA 1 DE VAL DA LAJE (ALLUÉ, 2000); GSL = Nª Sª DAS LAPAS (ALLUÉ, 2000); SMAC = SANTA MARGARIDA DA COUTADA (FERREIRA, 2010).

first anthropic markers may go back to 6000 BP and, even though the development of scrubs of Erica dates back to 5000 BP (Mateus and Queiroz, 1993), species of mesophyll character are increasingly encountered since this phase.

In the Final Neolithic/Chalcolithic we can see the maximum values for the populations of Ericaceae. *Pinus* has a very low representativeness and deciduous *Quercus* presents a gradual decrease. *Vitis*, *Olea*, wild grasses and other grasses associated to farmlands present a higher representativeness; also, cereals appear in the palynological studies. Ericaceae (*Arbutus unedo*, *Erica arborea*, *Calluna*) remain the most representative taxa, but other shrubs (Cistaceae, *Pistacia*, Leguminosae, *Ligustrum*, and *Myrtus*) were identified (Figueiral, 1998; Allué, 2000).

3.3. Zooarchaeology

Epipalaeolithic/Mesolithic archaeofauna is registered in regions west of Alto Ribatejo (Detry, 2007, Valente, 2008). Regional Neolithic zooarchaeological data is quite sparse and circumscribed to the Limestone Massif (Almeida *et al.*, 2014; Valente and Carvalho, 2014). Early Neolithic data is to be found in Caldeirão, Pena d'Água and Cerradinho do Ginete (table 3). In the first, *Sus scrofa* predominates in terms of NISP values (Rowley-Conwy, 1992; cf. Davis, 2002; Albarella *et al.*, 2005), but *Bos taurus* and *Ovis aries* (as well as *Ovis aries/Capra hircus*) are also present. Pena d'Água presents a predominance of *Cervus elaphus*, followed by *Ovis aries/Capra hircus*, *Bos* sp. and *Sus* sp.

Middle Neolithic data is to be found in Pena d'Água, Costa do Pereiro and Cadaval where a predominance of *Cervus elaphus* or *Ovis aries/Capra hircus* is seen; other taxa (i.e., *Bos* sp., *Sus* sp.) have vestigial values. As for the Final Neolithic, regional data concerns only Cadaval and Morgado superior, but chronological revisions are on-going. While *Sus* sp. remains difficult to specifically

Taxon	Early Neolithic	Middle Neolithic	Final Neolithic
Bos sp.	++	+	+
Cervus elaphus	++	+++	+
Sus sp.	+++	++	+
Capreolus capreolus	+	+	+
Ovis aries/Capra hircus	++	+++	++

TABLE 3. GENERAL ABUNDANCE OF THE MAIN TAXA IDENTIFIED IN THE ALTO RIBATEJO ZOOARCHAEOLOGICAL RECORDS BASED ON NISP VALUES (+ <10; ++ 10-50; +++ >50). EARLY NEOLITHIC: CALDEIRÃO (ROWLEY-CONWY, 1992), PENA D'ÁGUA (VALENTE, 1998; CARVALHO ET AL., 2004; CARVALHO, 2008), CERRADINHO DO GINETE (CARVALHO ET AL., 2004); MIDDLE NEOLITHIC: PENA D'ÁGUA (VALENTE, 1998; CARVALHO, 2008; LUÍS ET AL., 2013), COSTA DO PEREIRO (CARVALHO, 2008), CADAVAL (ALMEIDA ET AL., IN PRESS). FINAL NEOLITHIC: CADAVAL AND MORGADO SUPERIOR (ALMEIDA ET AL., IN PRESS).

identify (e.g., Albarella et al., 2005), Bos sp. corresponds to the domestic variant whenever a specific adscription is possible, exception being one tooth identified in the Early Neolithic of Pena d'Água, with some reservations (Carvalho, 2008; Valente and Carvalho, 2014).

3.4. Rock art and other manifestations

In the western areas of the Alto Ribatejo, symbolic evidences correspond mainly to portable features, as schist geometrically engraved plaquettes and shell or stone ornaments (bracelets, beads, pendants). An example is the Carrascos cave, where personal adornment objects represent a high percentage of the material found; or the Galinha cave, where interesting schist plaquettes were found, along with other Neolithic materials (Gonçalves and Pereira, 1974/1977). In contrast, no evident purely 'symbolic' items are found in the eastern areas, even in the first megalithic burials (Oosterbeek et al., 2014). Here, symbolic representations are concentrated in the Tagus Valley Rock Art Complex, where several rock art sites (Serrão, 1972) occur substantially from the Cedilho dam in Spain to the mouth of the Ocreza River (Garcês, 2009), reaching some tributary rivers. The typology of engravings is very diverse: zoomorphic, anthropomorphic and a majority of geometric figures (e.g., circles, concentric circles, lines, cup-marks).

During the last few years there is a tendency to characterize some motifs that come across the Iberian Peninsula that do not exactly fit into the naturalism of Palaeolithic rock art neither on the Neolithic schematic features (Collado-Giraldo, 2004, 2006; Bueno-Ramírez et al., 2009). Collado proposes an integration of these figures on a new rock art artistic cycle denominated *pre-schematic rock art* (Collado-Giraldo, 2004, 2006). This *style* breaks with the traits of Magdalenian filiform features and is characterized by the almost exclusive use of pecking but it also emerges figurative painted elements (Collado-Giraldo and García-Arranz, 2006).

The Tagus valley presents a wide chronology where evidences increasingly show us a sequence between Palaeolithic figures (Baptista, 2001), the figures of the Zêzere valley and post-Palaeolithic figures. These pre-schematic features could be identified due to figures like the deer in rock number 155 from Fratel. The motifs present in this rock are mainly big figures of animals with a diverse interior body decoration (x-rays and lines). All male deer antlers are found in perspective and, in spite of the body being displayed in profile, in six of these animal representations the front and back legs are also in perspective. Different ways of animal representation are patent: large animals profusely engraved and decorated internally with reticulated lines, wide necks and well-developed antlers or manes, and cervical-dorsal lines well pronounced, representing a naturalism that follows previous periods (M. V. Gomes sub-naturalism period); animals represented so much simpler than the previous, less obvious body lines (emerge with a more oval shape), interior decoration summarized only by

Taxon	Epipalaeolithic			Early Neolithic	Middle Neolithic				Late Neolithic	Chalcolithic		
	APA*	Amr*	PBoq^	APA*	PBoq^	VL^*	APA*	GSL*	APA*	PBoq	VL^*	SMAC^
Pinus sp.		+++	++		++	+				+	+	
Deciduous Quercus			++		++	+	+		+	+		+
Evergreen Quercus	+	+	++	+	++	+	+	++	+	++	+	+
Olea europaea	+++			+++		+	+++	+	++	+	+	
Riparian woodland			+++		++				+	++	+	
Ericaceae						+					+++	+
Arbutus unedo		+				++		+	+		+	
Erica sp.			+		++	+	+		+	++	++	
Cistaceae	+	+	+		+	+				+	+	
Rhamnus alaternus/ Phillyrea	+			+		+		+	+		+	
Vitis			+		+							
Poaceae			++		+++					+++	+	++
Asteraceae			++		++	++				+	+	++
Plantago sp.					+					+		+
Cerealia – type					+							+++

TABLE 2. GENERAL ABUNDANCE OF THE MAIN TAXA IDENTIFIED IN THE ALTO RIBATEJO PALYNOLOGICAL(^) AND ANTHRACOLOGICAL(*) RECORDS (+ 0-25%; ++ 25-50%; +++ >50%). APA = PENA D'ÁGUA (FIGUEIRAL, 1998); AMR = AMOREIRA (ALLUÉ 2000; FERREIRA ET AL., 2014); PBOQ = PAÚL DO BOQUILOBO (VIS ET AL., 2010); VL = ANTA 1 DE VAL DA LAJE (ALLUÉ, 2000); GSL = Nª Sª DAS LAPAS (ALLUÉ, 2000); SMAC = SANTA MARGARIDA DA COUTADA (FERREIRA, 2010).

first anthropic markers may go back to 6000 BP and, even though the development of scrubs of Erica dates back to 5000 BP (Mateus and Queiroz, 1993), species of mesophyll character are increasingly encountered since this phase.

In the Final Neolithic/Chalcolithic we can see the maximum values for the populations of Ericaceae. *Pinus* has a very low representativeness and deciduous *Quercus* presents a gradual decrease. *Vitis*, *Olea*, wild grasses and other grasses associated to farmlands present a higher representativeness; also, cereals appear in the palynological studies. Ericaceae (*Arbutus unedo*, *Erica arborea*, *Calluna*) remain the most representative taxa, but other shrubs (Cistaceae, *Pistacia*, Leguminosae, *Ligustrum*, and *Myrtus*) were identified (Figueiral, 1998; Allué, 2000).

3.3. Zooarchaeology

Epipalaeolithic/Mesolithic archaeofauna is registered in regions west of Alto Ribatejo (Detry, 2007, Valente, 2008). Regional Neolithic zooarchaeological data is quite sparse and circumscribed to the Limestone Massif (Almeida *et al.*, 2014; Valente and Carvalho, 2014). Early Neolithic data is to be found in Caldeirão, Pena d'Água and Cerradinho do Ginete (table 3). In the first, *Sus scrofa* predominates in terms of NISP values (Rowley-Conwy, 1992; cf. Davis, 2002; Albarella *et al.*, 2005), but *Bos taurus* and *Ovis aries* (as well as *Ovis aries*/*Capra hircus*) are also present. Pena d'Água presents a predominance of *Cervus elaphus*, followed by *Ovis aries*/*Capra hircus*, *Bos* sp. and *Sus* sp.

Middle Neolithic data is to be found in Pena d'Água, Costa do Pereiro and Cadaval where a predominance of *Cervus elaphus* or *Ovis aries*/*Capra hircus* is seen; other taxa (i.e., *Bos* sp., *Sus* sp.) have vestigial values. As for the Final Neolithic, regional data concerns only Cadaval and Morgado superior, but chronological revisions are on-going. While *Sus* sp. remains difficult to specifically

Taxon	Early Neolithic	Middle Neolithic	Final Neolithic
Bos sp.	++	+	+
Cervus elaphus	++	+++	+
Sus sp.	+++	++	+
Capreolus capreolus	+	+	+
Ovis aries/Capra hircus	++	+++	++

TABLE 3. GENERAL ABUNDANCE OF THE MAIN TAXA IDENTIFIED IN THE ALTO RIBATEJO ZOOARCHAEOLOGICAL RECORDS BASED ON NISP VALUES (+ <10; ++ 10-50; +++ >50). EARLY NEOLITHIC: CALDEIRÃO (ROWLEY-CONWY, 1992), PENA D'ÁGUA (VALENTE, 1998; CARVALHO ET AL., 2004; CARVALHO, 2008), CERRADINHO DO GINETE (CARVALHO ET AL., 2004); MIDDLE NEOLITHIC: PENA D'ÁGUA (VALENTE, 1998; CARVALHO, 2008; LUÍS ET AL., 2013), COSTA DO PEREIRO (CARVALHO, 2008), CADAVAL (ALMEIDA ET AL., IN PRESS). FINAL NEOLITHIC: CADAVAL AND MORGADO SUPERIOR (ALMEIDA ET AL., IN PRESS).

identify (e.g., Albarella *et al.*, 2005), *Bos* sp. corresponds to the domestic variant whenever a specific adscription is possible, exception being one tooth identified in the Early Neolithic of Pena d'Água, with some reservations (Carvalho, 2008; Valente and Carvalho, 2014).

3.4. Rock art and other manifestations

In the western areas of the Alto Ribatejo, symbolic evidences correspond mainly to portable features, as schist geometrically engraved plaquettes and shell or stone ornaments (bracelets, beads, pendants). An example is the Carrascos cave, where personal adornment objects represent a high percentage of the material found; or the Galinha cave, where interesting schist plaquettes were found, along with other Neolithic materials (Gonçalves and Pereira, 1974/1977). In contrast, no evident purely 'symbolic' items are found in the eastern areas, even in the first megalithic burials (Oosterbeek *et al.*, 2014). Here, symbolic representations are concentrated in the Tagus Valley Rock Art Complex, where several rock art sites (Serrão, 1972) occur substantially from the Cedilho dam in Spain to the mouth of the Ocreza River (Garcês, 2009), reaching some tributary rivers. The typology of engravings is very diverse: zoomorphic, anthropomorphic and a majority of geometric figures (e.g., circles, concentric circles, lines, cup-marks).

During the last few years there is a tendency to characterize some motifs that come across the Iberian Peninsula that do not exactly fit into the naturalism of Palaeolithic rock art neither on the Neolithic schematic features (Collado-Giraldo, 2004, 2006; Bueno-Ramírez *et al.*, 2009). Collado proposes an integration of these figures on a new rock art artistic cycle denominated *pre-schematic rock art* (Collado-Giraldo, 2004, 2006). This *style* breaks with the traits of Magdalenian filiform features and is characterized by the almost exclusive use of pecking but it also emerges figurative painted elements (Collado-Giraldo and García-Arranz, 2006).

The Tagus valley presents a wide chronology where evidences increasingly show us a sequence between Palaeolithic figures (Baptista, 2001), the figures of the Zêzere valley and post-Palaeolithic figures. These pre-schematic features could be identified due to figures like the deer in rock number 155 from Fratel. The motifs present in this rock are mainly big figures of animals with a diverse interior body decoration (x-rays and lines). All male deer antlers are found in perspective and, in spite of the body being displayed in profile, in six of these animal representations the front and back legs are also in perspective. Different ways of animal representation are patent: large animals profusely engraved and decorated internally with reticulated lines, wide necks and well-developed antlers or manes, and cervical-dorsal lines well pronounced, representing a naturalism that follows previous periods (M. V. Gomes sub-naturalism period); animals represented so much simpler than the previous, less obvious body lines (emerge with a more oval shape), interior decoration summarized only by

one or few transversal lines (the so-called *line-of-life*), also with antlers (in the case of male cervids) in perspective, as well as his back and front legs in some cases (M. V. Gomes transition period from the Neolithic to Epipalaeolithic – period II stylized-static) (Gomes, 2007, 2010).

4. Final considerations

At the current state of knowledge, the earliest neolithisation of the Alto Ribatejo seems to have occurred in straight connection to the western coastal dynamics (Zilhão, 2001; Carvalho, 2008). Nonetheless, as foreseen in the 90s (Oosterbeek, 1994), some aspects related to the material culture and symbolic behaviour, among others, evidenced in the eastern areas, seem to relate to an important influence of other inland regions in the beginning and development of the Alto Ribatejo Neolithic.

This raises some problems, namely the abovementioned Epipalaeolithic and Mesolithic regional low archaeological visibility. Sites located in the Limestone Massif (Bicho *et al.*, 2003; Carvalho, 2008) and in the Tagus Basin (Oosterbeek, 1994; Cruz, 1997, 2011), although scarce and not without problems, represent the sparse regional Holocene hunter-gathers. We believe that this scenario might also relate to the focus of archaeological research paid to the karstic areas, with a lack of field surveys in the inland. Also, as is indicated for the contexts of the latter areas, the predominance of quartzite post-glacial macrolithic industries might have historically biased the recognition of Epipalaeolithic and Mesolithic occupations – this is something that future research should address. Recent studies in Portugal and Spain (e.g., Neves *et al.*, 2008; Arias *et al.*, 2009; Cerrillo *et al.*, 2014), following updated methodologies, allowed for the recognition of several occupations ranging from the Epipalaeolithic to the Middle Neolithic in areas thought to have none or little occupation in these periods. Also, the chronological and geographical gaps between the initial moments of the Early Neolithic are fading. The review of Holocene macrolithic collections of the Alto Ribatejo also points in this direction (Rosina *et al.* 2010).

Another interesting aspect that future research should focus is the acquisition of data related to possible agricultural practices in the Neolithic of Western Iberia. As discussed elsewhere, we still lack data regarding this important aspect of the 'Neolithic way of life', even if a few direct or indirect evidences exists for the Lower Tagus Valley (López-Dóriga and Simões, 2012; Carvalho *et al.*, 2013). This, together with other still not largely addressed possibilities (e.g., isotopic studies, aDNA), should be an opportunity for the acquisition of relevant data.

Aknowledgements

Nelson J. Almeida (SFRH/BD/78079/2011), Cristiana Ferreira (SFRH/BD/78542/2011) and Sara Garcês (SFRH/BD/69625/2010) benefit from FCT PhD individual scholarships under QREN – POPH – Typology 4.1. – Advanced Training subsidized by the European Social Fund and by national MEC funds.

References

Albarella, U.; Davis, S. J.; Detry, C.; Rowley-Conwy, P., 2005 – Pigs of the 'Far West': the biometry of *Sus* from archaeological sites in Portugal. Anthropozoologica 40:2, pp. 27-54.

Allué, E., 2000 – Pollen and charcoal analysis from archaeological sites from the Alto Ribatejo (Portugal). In: Cruz, A. R.; Oosterbeek, L., eds.- Territórios, Mobilidade e Povoamento no Alto Ribatejo: Indústrias e Ambientes. Arkeos 9. Tomar: Ceiphar, pp. 37-57.

Almeida, N. J.; Ferreira, C.; Allué, E.; Burjachs, F.; Cruz, A. R.; Oosterbeek, L.; Rosina, P.; Saladié, P., 2014 – Acerca do impacte climático e antropozoogénico nos inícios da economia produtora: o registo do Alto Ribatejo (Portugal Central, Oeste Ibérico). In: Zocche, J.; Campos, J. B.; Almeida, N. J.; Ricken, C., orgs.- Arqueofauna e Paisagem. Erichim: Habilis Editora, pp. 63-84.

Almeida, N. J.; Saladié, P.; Oosterbeek, L., in press – Zooarqueologia e Tafonomia da Gruta da Nossa Senhora das Lapas e Gruta do Cadaval (Alto Ribatejo, Portugal Central). In: Actas do 5º Congresso do Neolítico Peninsular, Lisboa, 07-09 de Abril de 2011.

Arias, P.; Cerrillo Cuenca, E.; Álvarez Fernández, E.; Gómez-Pellón, E.; González Cordero, A., 2009 – A view from the edges: the Mesolithic settlement of the interior areas of the Iberian Peninsula reconsidered. In: McCartan, S.; Schulting, R.; Warren, G.; Woodman, P., eds.- Mesolithic horizons. Conference on the Mesolithic in Europe, Belfast. Vol. I, pp. 303-311.

Arnaud, J. M., 1982 – Néolithique Ancien et processus de Néolithisation dans le Sud du Portugal. In: Archéologie en Languedoc, N.º Spécial, Actes du Colloque International de Préhistoire. pp. 29-48.

Baptista, A. M., 2001 – Ocreza (Envendos, Mação, Portugal central): um novo sítio com arte paleolítica de ar livre. In: Cruz, A. R.; Oosterbeek, L., coord.- Territórios, mobilidade e povoamento no Alto-Ribatejo. II: Santa Cita e o quaternário da região. Arkeos 11. Tomar: Ceiphar, pp. 163-192.

Bicho, N. F.; Haws, J.; Hockett, B.; Markova, A.; Belcher, W., 2003 – Paleoecologia e ocupação humana da Lapa do Picareiro: resultados preliminares. Revista Portuguesa de Arqueologia. 6(2), pp. 49-81.

Bronk Ramsey, C., 2009 – Bayesian analysis of radiocarbon dates. Radiocarbon. 51(1), pp. 337-360.

Bueno-Ramírez, P.; Balbín-Behrmann, R.; Alcolea-González, 2009 – Estilo V en el âmbito del Duero: cazadores finiglaciares en Siega Verde (Salamanca). Arte Prehistórico al aire libre en el Sur de Europa. pp. 259-490.

Burbidge, C. I.; Trindade, M. J.; Dias, M. I.; Oosterbeek, L.; Scarre, C.; Rosina, P.; Cruz, A.; Cura, S.; Cura, P.; Caron, L.; Prudêncio, M. I.; Cardoso, G. J. O.; Franco, D.; Marques, R.; Gomes, H., 2014 – Luminescence dating and associated analyses in transition landscapes of the Alto Ribatejo, central Portugal. Quaternary Geochronology. 20, pp. 65-77.

Burjachs, F., 1990 – Palinologia dels Dólmens de l'Alt Empordà i dels diposits quaternaris de la Cova de l'Arbreda (Serinyà, Pla de l'Estany) i del Pla de l'Estany (Olot, Garrotxa). Evolució del paisatge vegetal i del clima de fa més de 140.000 anys a N.E. de la Península Ibèrica. PhD Thesis. Universitat Autònoma de Barcelona (Spain). Published in microfilm (ISBN:84.7488.783.2).

Burjachs, F.; Allué, E., 2003 – Paleoclimatic evolution during the Last Glacial Cycle at the NE of the Iberian Peninsula. In: Ruiz Zapata, M. B.; Dorado Valiño, M.; Valdeolmillos, A.; Gil García, Bardají Azcárate, M. J.; Bustamante, I.; Martínez, I., eds.- Quaternary Climatic Changes and Environmental Crises in the Mediterranean Region. Universidad de Alcalá: Madrid, pp. 191-200.

Carvalho, A. F., 1998 – O Abrigo da Pena d'Água (Rexaldia, Torres Novas): resultados dos trabalhos de 1992-1997. Revista Portuguesa de Arqueologia. 1(2), pp. 39-72.

Carvalho, A. F., 2008 – A Neolitização do Portugal Meridional. Os exemplos dos Maciço Calcário Estremenho e do Algarve Ocidental. Promontoria Monográfica 12. Faro: Universidade do Algarve. 428 p.

Carvalho, A. F.; Valente, M. J.; Haws, J. A., 2004 – Faunas mamalógicas do Neolítico antigo do Maciço Calcário Estremenho: análise preliminar de dados recentes. Promontoria. 2(2), pp. 143-155.

Carvalho, A. F., Gibaja, J. F., Cardoso, J. L., 2013 – Insights into the earliest agriculture of Central Portugal: sickle implements from the Early Neolithic site of Cortiçóis (Santarém). Comptes Rendus Palevol. 12(1), pp. 31-43.

Cerrillo Cuenca, E.; González Cordero, A., 2014 – Collective burial caves in spanish Extremadura: chronology, landscapes and identities. In: Cruz, A.; Cerrillo Cuenca, E.; Bueno Ramírez, P.; Canincas, J. C.; Batata, C., eds.- Rendering Death: Ideological and Archaeological Narratives from Recent Prehistory (Iberia), Proceedings of the conference held in Abrantes, Portugal, 11 May 2013. BAR International Series 2648. Oxford: Archaeopress, pp. 77-89.

Collado Giraldo, H., 2004 – Un nuevo ciclo de arte prehistórico en Extremadura: el arte rupestre de las sociedades de economía cazadora recolectora durante el Holoceno Inicial como precedente del arte rupestre esquemático en Extremadura. In: Calado, M., ed.- Sinais de Pedra. Actas del 1º Colóquio Internacional sobre Megalitismo e Arte Rupestre na Europa Atlântica. Évora: Fundação Eugénio de Almeida (CD ROM).

Collao Giraldo, H., 2006 – Arte Rupestre en la Cuenca del Guadiana: El conjunto de grabados del Molino Manzánez (Alconchel-Cheles, Badajoz). Memorias d'Odiana. Estudos Arqueológicos do Alqueva, N.º 4. Beja: EDIA.

Collado Giraldo, H.; García Arranz, J. J., coord., 2006 – Arte rupestre en el Parque Natural de Monfragüe: el Sector Oriental (Corpus de Arte Rupestre en Extremadura, vol. I). Mérida: Junda de Extrremadura, Consejería de Cultura.

Cruz, A. R., 1997 – Vale do Nabão: do Neolítico à Idade do Bronze. Arkeos 3. Tomar: Ceiphar. 361 p.

Cruz, A. R., 2011 – A Pré-História recente no Vale do Baixo Zêzere. Um olhar diacrónico. PhD Thesis. Universidade de Trás-os-Montes e Alto Douro (Portugal).

Cruz, A. R., 2013 – The settlement of Fontes and Pedra da Encavalada – key sites for the Neolithization of North Ribatejo (Portugal). In: Cruz, A. R.; Graça, A.; Oosterbeek, L.; Rosina, P., eds.- Iº Congresso de Arqueologia do Alto Ribatejo. Arkeos 34. Tomar: Ceiphar, pp. 79-93.

Cruz, A. R., in press – Transition into the Production Process in Middle Tagus (Central Portugal). Proceedings of the XVII UISPP Conference, Burgos, 2014, Session B20 Contexts without definition, definitions without contexto. Arguments for the characterization of the (Pre)historic realities during the Neolithisation of the western mediterranean.

Cruz, A. R.; Grimaldi, S.; Oosterbeek, L.; Rosina, P., 2000 – Indústrias macrolíticas do pós-glaciar no Alto Ribatejo. In: Cruz, A. R.; Oosterbeek, L., eds.- Territórios, Mobilidade e Povoamento no Alto Ribatejo I: Indústrias e Ambientes. Arkeos 9. Tomar; Ceiphar, pp. 9-21.

Davis, S. J. M., 2002 – The mammals and birds from Gruta do Caldeirão, Portugal. Revista Portuguesa de Arqueologia. 5 (2), pp. 29-98.

Detry, C., 2007 – Paleoecologia e Paleoeconomia do Baixo Tejo no Mesolítico Final: o contributo do estudo dos mamíferos dos concheiros de Muge. PhD Thesis. Universidad de Salamanca (Spain)/ Universidade Autónoma de Lisboa (Portugal).

Diniz, M., 2007 – O Sítio da Valada do Mato (Évora): aspectos da neolitização no interior/sul Portugal. In Trabalhos de Arqueologia 48. Lisboa: Instituto Português do Património Arquitectónico e Arqueológico.

Ferreira, C., 2010 – Contribuição para o Estudo das Transformações Ambientais na Transição para o Agro-Pastoralismo no Alto Ribatejo. Master Thesis. Universidade de Trás-os-Montes e Alto Douro, Instituto Politécnico de Tomar (Portugal).

Ferreira, C.; Allué, E.; Cruz, A.; Rosina, P.; Oosterbeek, L., 2014 – Estudo Antracológico do Povoado Neolítico da Amoreira (Alto Ribatejo, Centro de Portugal). In: Actas das IX Jornadas de Arqueologia Ibero-americanas e I Jornadas de Arqueologia Transatlântica. Criciúma: Habilis Editora, pp. 41-50.

Figueiral, I., 1998 – O Abrigo da Pena d'Água (Torres Novas): a contribuição da antracologia. Revista Portuguesa de Arqueologia. 1(2), pp. 73-79.

Garcês, S., 2009 – Cervídeos na Arte Rupestre do Vale do Tejo. Contributo para o estudo da Pré-história Recente. Master Thesis. Universidade de Trás-os-Montes e Alto Douro, Instituto Politécnico de Tomar (Portugal).

Grau, E.; Duque, D., 2007 – Los Paisajes Rurales Protohistóricos: una síntesis arqueobotánica. In: Rodríguez Díaz, A.; Pavón Soldevila, I., coord.- Arqueología de la tierra: paisajes rurales de la protohistoria peninsular: VI cursos de verano internacionales de la Universidad de Extremadura (Castuera, 5-8 de julio de 2005). Universidad de Extremadura, pp. 297-326.

Gonçalves, V. M. S.; Pereira, A. R., 1974/1977 – Considerações sobre o espólio neolítico da Gruta dos Carrascos. O Arqueólogo Português. Lisboa. 3ª série: 79, pp. 49-87.

Gomes, M. V., 2007 – Os períodos iniciais da arte do Vale do Tejo (Paleolítico e Epipaleolítico). Cuadernos de Arte Rupestre. 4, pp. 81-116.

Gomes, M. V., 2010 – Um Ciclo Artístico-Cultural Pré e Proto-Histórico. PhD Thesis. Universidade Nova de Lisboa (Portugal). 2 vols.

Grimaldi, S.; Rosina, P.; Cruz, A. R.; Oosterbeek, L., 1999 – A geo-archaeological interpretation of some 'Languedocian' lithic collections of the Alto Ribatejo (Central Portugal). In: Cruz, A. E.; Milliken, S.; Oosterbeek, L.; Peretto, C., eds.- Human populations in the circum-Mediterranean area. Arkeos 5. Tomar: Ceiphar, pp. 231-241.

Guilaine, J.; Veiga Ferreira, O. da, 1970 – Le Néolithique ancien au Portugal. Bulletin de la Société préhistorique française. Tome 67, N.º 1, pp. 304-322.

Jalut, F.; Estebanamat, A.; Mora, S.; Fontugne, M.; Mook, R.; Bonnet, L.; Gauquelin. T., 1997 – Holocene climatic changes in the western Mediterranean: installation of the Mediterranean climate. Comptes Rendus de l'Académie des Sciences – Series IIA – Earth and Planetary Science. Vol. 325(5), pp. 327-334.

Lopez-Doriga, I.; Simões, T., 2012 – Early Neolithic agriculture in Portugal: the evidence from plant macro-remains. Poster presented in the Environmental Archaeologies of Neolithisation, 10-12 November 2012, University of Reading.

Luís, S.; Correia, F.; Fernandes, P. V., 2013 – Middle Neolithic Zooarchaeology at the Pena d'Água Rock-shelter (Portuguese Estremadura). In: VI Jornadas de Jóvenes en Investigación Arqueológica, 7-11th May, 2013. Barcelona: Universitat de Barcelona.

Mateus, J. E.; Queiroz, P. F., 1993 – Os Estudos de Vegetação Quaternária em Portugal: contextos, balanço de resultados, perspectivas. In: Carvalho, G. S.; Ferreira, A. B.; Senna-Martinez, J. C., coord.- O Quaternário em Portugal, Balanço e Perspectivas. Lisboa: Colibri, pp. 105-131.

Neves, C.; Rodrigues, F.; Diniz, M., 2008 – Neolithisation process in lower Tagus valley left bank: old perspectives and new data. In: Diniz, M., ed.- Early Neolithic in the Iberian Peninsula: regional and transregional components. BAR International 1857. Oxford: Archaeopress, pp. 43-51.

Oosterbeek, L., 1994 – Echoes from the East: the western network. An insight to unequal and combined development, 7000-2000 BC. PhD Thesis. University of London (England).

Oosterbeek, L.; Almeida, N.; Garcês, S., 2014 – Territories revisited: identities and exclusion as seen from an archaeological perspective. In: Lins, M.; Borges, S.; Oosterbeek, L.; Mendes, A.; Leite, D.; Lima, A., eds.- Identidades e diversidade cultural. Etnia e género. Teresina: Fundação Quixote – CEIPHAR/ITM, pp. 65-77.

Reimer, P. J.; Bard, E.; Bayliss, A.; Beck, J. W.; Blackwell, P. G.; Bronk Ramsey, C.; Grootes, P. M.; Guilderson, T. P.; Haflidason, H.; Hajdas, I.; Hatté, C.; Heaton, T. J.; Hoffman, D. L.; Hogg, A. G.; Hughen, K. A.; Kaiser, K. F.; Fromer, B.; Manning, S. W.; Niu, M.; Reimer, R. W.; Richards, D. A.; Scott, E. M.; Southon, J. R.; Staff, R. A.; Turney, C. S. M.; Van der Plicht, J., 2013 – IntCal13 and Marine13 Radiocarbon Age Calibration Curves 0-50,000 Years cal BP. Radiocarbon. 55(4), pp. 1869-1887.

Rosina, P.; Cura, S.; Oosterbeek, L.; Grimaldi, S.; Cruz, A.; Gomes, J., 2010 – Crono-estratigrafia das ocupações humanas quaternárias do Alto Ribatejo e a problemática dos complexos macrolíticos. In: Materiaes para o estudo das antiguidades portuguesas – número especial, Castelo Branco, pp. 107-148.

Rowley-Conwy, P., 1992 – The early Neolithic animal bones from Gruta do Caldeirão. In: Zilhão, J., ed.- Gruta do Caldeirão. O Neolítico Antigo. Trabalhos de Arqueologia 6. Lisboa: Instituto Português do Património Arquitectónico e Arqueológico, pp. 231-257.

Serrão, E. C., 1972 – O Complexo de Arte Rupestre do Tejo (Vila Velha de Ródão-Nisa). Notícia preliminar. Separata de Arqueologia e História. 9ª série, Vol. IV.

Scarre, C.; Oosterbeek, L., 2010 – The megalithic tombs of the middle Tagus basin and agro-pastoral origins in Western Iberia. In: Armbruester, T.; Hegewisch, M., eds.- Beiträge zur Vor- und Frühgeschichte der Iberischen Halbinsel und Mitteleuropas. Studien in honorem Philine Kalb. Habelt: Bonn, pp. 97-110.

Silva, J. C.; Batista, Á.; Gaspar, F., 2009 – Carta Arqueológica do Concelho de Abrantes. Câmara Municipal de Abrantes (CD ROM).

Valente, M. J., 1998 – Análise preliminar da fauna mamalógica do Abrigo da Pena d'Água (Torres Novas): campanhas de 1992-1994. Revista Portuguesa de Arqueologia. 1(2), pp. 85-96.

Valente, M. J., 2008 – As últimas sociedades de caçadores-recolectores no Centro e Sul de Portugal (10.000-6.000 anos BP): aproveitamento dos recursos animais. PhD Thesis. Universidade do Algarve (Portugal).

Valente, M. J.; Carvalho, A. F., 2014 – Zooarchaeology in the Neolithic and Chalcolithic of Southern Portugal. Environmental Archaeology. 19(3), pp. 226-240.

Vis, G.-J.; Bohncke, S.; Schneider, H.; Kasse, C.; Coenraads-Nederveen, S.; Zuurbier, K.; Rozema, J., 2010 – Holocene flooding history of the Lower Tagus Valley (Portugal). Journal of Quaternary Science. 25(8), pp. 1222-1238.

Zilhão, J., ed. 1992 – Gruta do Caldeirão. O Neolítico Antigo. Trabalhos de Arqueologia 6. Lisboa: Instituto Português do Património Arquitectónico e Arqueológico.

Zilhão, J., 1993 – The Spread of Agro-pastoral Economies across Mediterranean Europe: view from the far west. Journal of Mediterranean Archaeology. 6, pp. 5-63.

Zilhão, J., 2001 – Radiocarbon evidence for maritime pioneer colonization at the origins of farming in west Mediterranean Europe. Proceedings of the National Academy of Sciences USA. 24(98), pp. 14180-14185.

Mobility in late Prehistory in Galicia: a preliminary interpretation from pottery

M. Pilar Prieto Martínez (pilar.prieto@usc.es),
and Óscar Lantes Suárez (oscar.lantes@usc.es)
University of Santiago de Compostela (Spain)

Abstract

The north-west of the Iberian Peninsula has traditionally been considered as a key region for the connection of remote parts of the European continent. However, very little research has been carried out in detail on the available archaeological record, aimed at verifying this apparently obvious hypothesis. This idea was mainly supported by studying metals, especially from the later stages of the Bronze Age, in order to justify this importance.

We believe that if a detailed study is carried out of other elements of material culture, it could demonstrate the importance of the geographical location in a more specific manner. This is the case of this study, in which we present specific cases of pottery in which it is possible to document foreign influences in different scales of distance, thereby demonstrating the frequent mobility of people in the prehistory of this region.

Our aim is to attempt to verify this hypothesis based on a study of pottery using traditional archaeological research methods (the typology and operational sequence), and archaeometric methods (based on the physical and chemical analysis of clays).

Through the typological study, based on the manufacturing methods (the type of temper used) and the decorative techniques (such as the use of shells, the presence of stab-and-drag or 'boquique' decoration, sgraffito or stamping), it is possible to document the influence of foreign traditions on prehistoric Galician pottery, as well as its possible points of origin. These influences can then be explored in greater detail through archaeometry. Studying the mineralogy of the pottery (using X-ray diffraction) in contrast with the mineralogical offer of the study areas has provided us with information on the most likely distances involved in providing the raw materials.

Therefore, based on the typological studies and analytical results for the pottery, we will cover a period of some four thousand years, from the Early Neolithic through to the First Iron Age, although the best-documented stage is associated with Bell Beaker contexts.

Key words: *Neolithic, Bronze Age, stab-and-drag or 'boquique' decoration, Penha type pottery, bell beaker pottery, wide horizontal rim vessel (WHR), NW Iberian Peninsula. Archaeometry, XRD, Geological Study, Provenance study, Pottery provenance*

1. Introduction

Traditionally, the north west of the Iberian Peninsula has been considered a key territory in terms of connections between very distant parts of the European continent, an idea that has mainly been supported through the study of metals.

The mobility of populations all over the planet has been demonstrated from their very earliest origins. This mobility occurred in very different ways, with people transporting objects and/or raw materials and transmitting ideas. The mechanisms of circulation vary depending on the region, period or culture being studied.

We have taken as our starting point the idea that the north-west Iberian Peninsula was an important area in terms of communication in Europe, that the mechanisms of circulation were equally important by land and by sea, and that these varied over time.

However, we are interested in explaining this from a different perspective to the one that is usually used: by studying pottery. Pottery has the potential of providing us with a large amount of information in a region where no organic remains are preserved.

Initially, we intend to characterise the mechanisms of circulation used throughout prehistory (from the Early Neolithic to the Late Bronze Age) based on a study of pottery, combining two types of approaches, through archaeology and archaeometry, with the aim of putting forward a hypothesis in relation to the intensity, frequency and even the direction of this mobility.

It may seem that it is impossible to discover the mobility of prehistoric societies through the archaeometry of pottery, because apart from analysing the pottery itself, it is necessary to analyse the sediments, with the difficulty of knowing which are the possible source areas out of all of those that exist, and which quarries were selected by prehistoric societies to make their pottery. A combined knowledge of the typology and technology allows us to identify and choose a selection of the most suitable sample for analysis. Combining this with Archaeometry offers great potential, at least in terms of confirming or ruling out whether the local lithology is coherent with the composition of the pottery. This aspect is a necessary initial step before continuing to carry out analyses which can refine these results. It is important to highlight the importance of the studies carried out on a wide territorial scale at analytical level (Salanova et al., 2015) and formal level (Prieto and Salanova, 2009; Prieto, 2012) in order to be able to consolidate and confirm the hypotheses made in this study.

2. Methodology and analytical process

In order to define similarities with other regions and possible areas of influence, archaeology mainly uses typology as a method, and to a lesser extent by studying the operative chain. Archaeometry is not always available as an approach, and when it is, in exceptional circumstances it can confirm the existence of foreign vessels in sites.

The pottery used in this study was analysed using different techniques, such as X-Ray Diffraction (XRD, to identify the mineralogy), X-Ray Fluorescence (XRF, to analyse majority elements and traces), CNSH analysis (carbon and nitrogen analysis), Solid Phase Colorimetry (to quantify the colours), Thin Section Analysis and surface analysis using binocular loupes (textural analysis), Pyrolysis-gas chromatography-mass spectrometry (to identify organic matter), Sr and Pb isotope analysis (precedence studies) and GIS analysis (to study the spatial distribution of the pottery). The methodology used in these analyses as well as the main results obtained are detailed in a number of published studies (Kaal et al., 2013; Lantes-Suárez et al., 2010, 2011; Martínez Cortizas et al., 2008, 2010, 2011; Prieto-Lamas et al., 2011; Prieto-Martínez et al., 2008, 2009A, 2009b, 2010).

In this study, we have used the mineralogical data obtained in DRX from crystalline powder. Using this data, it is possible to ascertain the types of mineralogy that are characteristic of the alteration material from different types of rocks (alkaline granites, calc-alkaline granites, acidic schists, basic schists or detritic deposits, predominated by quartz, amphibolites or ultrabasic rocks).

At the same time as identifying the mineralogy of the pottery, a study was carried out to identify the lithology of the area around the sites, in order to discover the available lithological and mineralogical offer. To do so, 1:50,000 scale geological maps from the Spanish Geological and Mining Institute (IGME) were used.

By comparing the composition of the pottery with the raw materials available in the surrounding area, it is possible to assign a series of minimum distances between the site and the most likely sources of raw material prima. The distances ranges established in this way are the following (Martínez Cortizas et al., 2011, Salanova et al., 2015):

- *in situ* (0-1 km),
- local (<7 km),

- district (7-50 km),
- regional (50-200 km)
- foreign (>200 km).

These distances ranges allow us to identify the areas of influence of a given settlement, or otherwise the sites from a given context.

3. Selected sites and vessels

We have made a selection of the pieces based on archaeological criteria (the possible external influence of the vessels) and archaeometric criteria (raw materials originating far away). These pieces (187) belong to 29 sites:[1] 17 settlements, 10 tombs and 2 ceremonial or unknown sites. Some sites have several prehistoric stages of occupation. It should be noted that the selection of pieces is not proportional, as 55% of them are from the Early Bronze Age, as this is the best-known period for the region, and for which we have the most samples that we can currently analyse.

4. Results

We have therefore made a summary that ranges from the Early Neolithic through to the Late Bronze Age.

Early Neolithic (6th millennium – 4500 BC)

Very few sites and pieces have been analysed from this period in the region, as to date only a small number have been found. We have only selected one piece, a fragment from the rim of a bowl with stab-and-drag and *boquique* decoration from one settlement (As Mamelas site) (fig. 1). The decorative

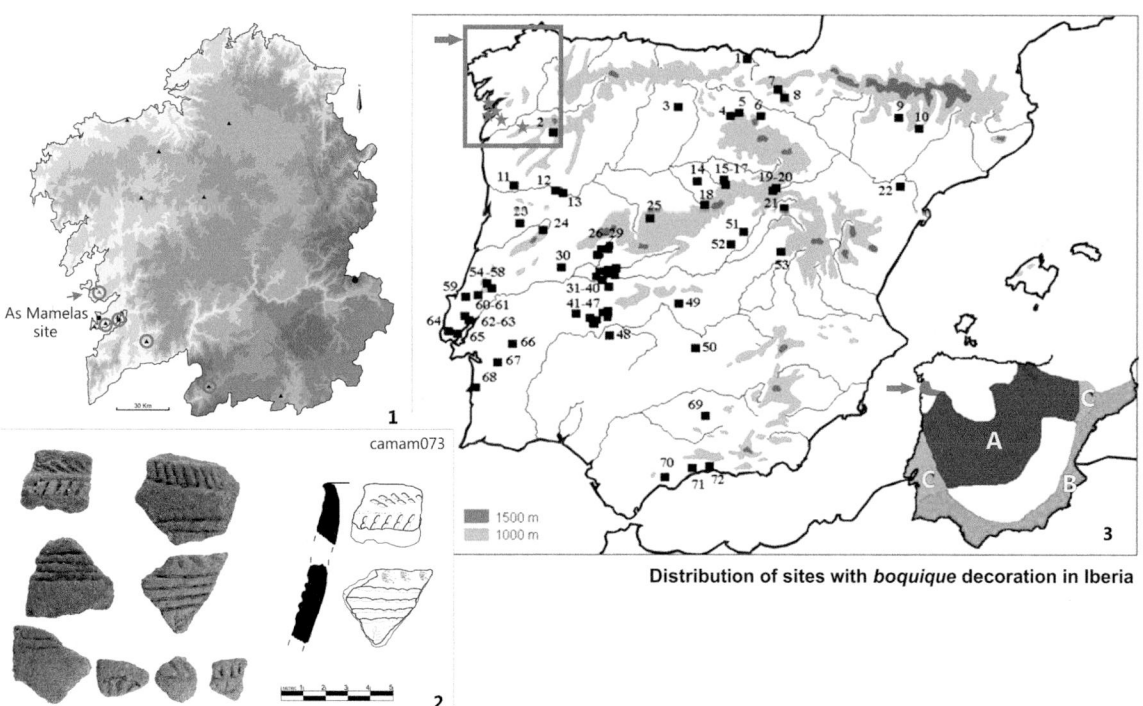

FIGURE 1. POTTERY ANALYSED FROM THE EARLY NEOLITHIC. 1. MAP SHOWING THE DISTRIBUTION OF THE EARLY NEOLITHIC SITES IN GALICIA, HIGHLIGHTING THE SITES WITH POTTERY WITH BOQUIQUE DECORATION (FROM PRIETO, 2010). 2. FRAGMENTS OF THE SAME VESSEL FROM THE SITE OF AS MAMELAS WITH BOQUIQUE DECORATION, ANALYSED IN THIS STUDY (ILLUSTRATION AND PHOTOS BY ERIC CARLSSON). 3. DISTRIBUTION OF AREAS WITH BOQUIQUE DECORATION FROM THE NEOLITHIC PERIOD IN THE IBERIAN PENINSULA (FROM ALDAY ET AL., 2009) TO WHICH WE HAVE ADDED THE GALICIAN AREA.

---

[1] Some of which are still unpublished.

	Granitic				Metamorphic				Amphibolic and Ultrabasic				
	Q	FK	Pla	Mic	Clo	Epi	Cli	Ana	Anf	Ser	Cal	Px	Tal
Early Neolithic (domestic contexts, N=1)													
ML073	23	13	41	12	5	–	–	–	6	–		–	–

	Alteration phyllosilicates						Iron oxides		
	Kao	Gib	Bay	Hal	Ver	Zeo	Hem	Goe	Ber
Early Neolithic (domestic contexts, N=1)									
ML073	–	–	–	–	–	–	–	–	–

TABLE 1A. MINEROLOGY OF THE POTTERY ANALYSED FROM THE EARLY NEOLITHIC (FOR ALL OF THE TABLES: Q. QUARTZ; FK: POTASSIUM FELDSPAR; PLA: PLAGIOCLASE; MIC: MICA; CLO: CHLORITE; EPI: EPIDOTE; CLI: CLINOZOISITE; ANA: ANATASE; ANF: AMPHIBOLE; SER: SERPENTINE; CAL: CALCITE, PX: PYROXENES; TAL: TALC; KAO: KAOLINITE; GIB: GIBBSITE; BAY: BAYERITE; HA: HALOISITE; VER: VERMICULITE; ZEO: ZEOLITE; HEM: HAEMATITE; GOE: GOETHITE; BER: BERNALITE).

Context	In situ	Local	District	Regional
	0-1 km	< 7 km	7-50 km	50-200 km
Domestic			1	

TABLE 1B. MINIMUM DISTANCES FROM THE SITE TO THE MOST LIKELY SOURCES OF RAW MATERIAL FOR THE POTTERY FROM THE EARLY NEOLITHIC.

technique has been documented in sites from the Iberian Peninsula in the sixth and fifth millennia BC. It is widely distributed, with two main concentrations in the Spanish region of Extremadura and in central Portugal (Alday *et al.*, 2009, Rojo *et al.*, 2012), and so far 5 sites are known in Galicia (Prieto, 2010).

The archaeometric results have made it possible to identify 6 minerals (table 1a). The most probably raw materials are basic schist with a possible granitic mixture and the most likely ceramic-source distance is between 7-50 km (within district range) (table 1b).

Based on the analytical and archaeological data, we believe that the piece was probably manufactured far from the site, but still in the NW Iberian Peninsula. As a result, we can identify mobility through the decorative technique. For the time being we cannot identify if the person who made the vessel came from another far-off location, or if they learned the technique from another potter.

Mid-Neolithic (4500 -3100/3000 BC)

We have selected 6 vessels from 3 sites in the region: Dombate Dolmen, Monte da Romea Barrow and Devesa do Rei site (fig. 2). This is only a small number of pieces, but they are representative of the types classified in this period (Prieto, 2004, 2009; Prieto *et al.*, 2012), in particular a carinated vessel that does not fit in with the regional style. Its angular profile and fine finish as well as its decoration made using a sea snail shell could have an external influence, probably from the Castellic Culture of Brittany (Cassen *et al.*, 2012).

The archaeometric results have made it possible to identify 10 minerals (table 2a). The most likely raw materials were alkaline granites in the domestic contexts, with a greater variety in the funerary contexts (alkaline granite, basic schist and amphibolite). The ceramic-source distances vary between a local range in the settlements and mainly district ranges in the tombs, including the carinated vessel from Dombate (table 2b).

Based on the available data, we believe that the greater mobility of the pottery from the tombs is coherent with communities who did not have a stable habitat, but who returned to their cemeteries

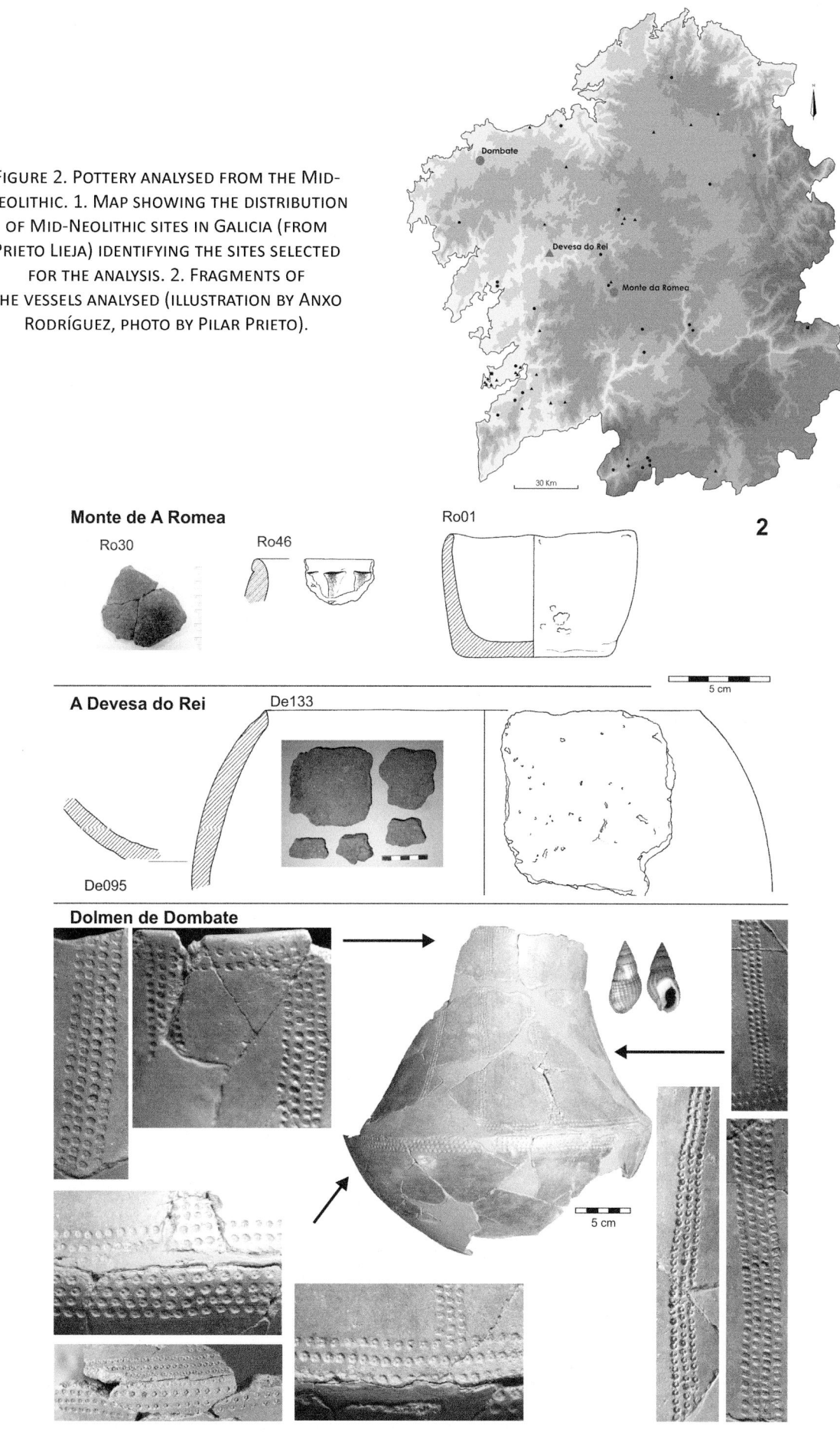

Figure 2. Pottery analysed from the Mid-Neolithic. 1. Map showing the distribution of Mid-Neolithic sites in Galicia (from Prieto Lieja) identifying the sites selected for the analysis. 2. Fragments of the vessels analysed (illustration by Anxo Rodríguez, photo by Pilar Prieto).

	Granitic				Metamorphic				Amphibolic and Ultrabasic				
	Q	FK	Pla	Mic	Clo	Epi	Cli	Ana	Anf	Ser	Cal	Px	Tal
Mid-Neolithic (Domestic contexts, N=2)													
De095	53	22	–	17	–	–	–	–	–	–	–	–	–
De133	39	48	8	5	–	–	–	–	–	–	–	–	–
Mid-Neolithic (funerary contexts, N=4)													
F%	100	75	100	75	50	–	–	25	75	–	–	–	–
Ab	31	4	24	6	11	–	–	2	24	–	–	–	–

	Alteration phyllosilicates						Iron oxides		
	Kao	Gib	Bay	Hal	Ver	Zeo	Hem	Goe	Ber
Mid-Neolithic (Domestic contexts, N=2)									
De095	–	–	–	8	–	–	–	–	–
De133	–	–	–	–	–	–	–	–	–
Mid-Neolithic (funerary contexts, N=4)									
F%	50	–	–	25	–	–	25	–	–
Ab	15	–	–	13	–	–	10	–	–

TABLE 2A. MINERALOGY OF THE POTTERY ANALYSED FROM THE MID-NEOLITHIC.

Context	In situ 0-1 km	Local < 7 km	District 7-50 km	Regional 50-200 km
Domestic		2		
Funerary	1		3	

TABLE 2B. MINIMUM DISTANCES FROM THE SITE TO THE MOST LIKELY SOURCES OF RAW MATERIAL FOR THE POTTERY FROM THE MID-NEOLITHIC.

in order to bury their dead. The carinated vessel from Dombate is included in this strategy, although its mineralogy is coherent with Galicia and Brittany, but the know-how behind its complicated manufacturing process is not Galician. This would seem to indicate that either the vessel was brought to the dolmen from a long way away, or that the potter who made it was foreign.

Late Neolithic (3100 -2400 BC)

36 vessels have been selected from 4 sites in the region: the barrow of Monte de Os Escurros and the settlements of Requeán, Montenegro and Zarra de Xoacín (fig. 3). The pieces are representative of the settlements, although it was practically impossible to analyse funerary pieces, which is why so few are shown. During this period, the distribution of pottery styles in the Iberian Peninsula is quite well defined. Galician pottery is included in the *Penha-type* tradition, distributed throughout its western half (Prieto, 2009). We can highlight 4 groups of pottery in this phase (Prieto, 2004):

1. Undecorated pottery.
2. Penha-type: pottery with typical Iberian designs: borders with combed decoration, metopes with reticulated designs or grooved vertical herringbone designs, incised triangles filled with printed dots. This type of pottery was first defined by Jorge (1986) for sites in the north of Portugal, and today it is often documented in the western half of the Iberian Peninsula as far as the Millares area.
3. Penha-type: anomalous profiles and decorations, but which fit within the Penha-type stylistic tradition.
4. Bell beaker imitations: pottery that combines aspects of pottery from Penha-type Chalcolithic traditions and bell beaker pottery. This is a recently characterised type (Prieto and Vázquez, 2011).

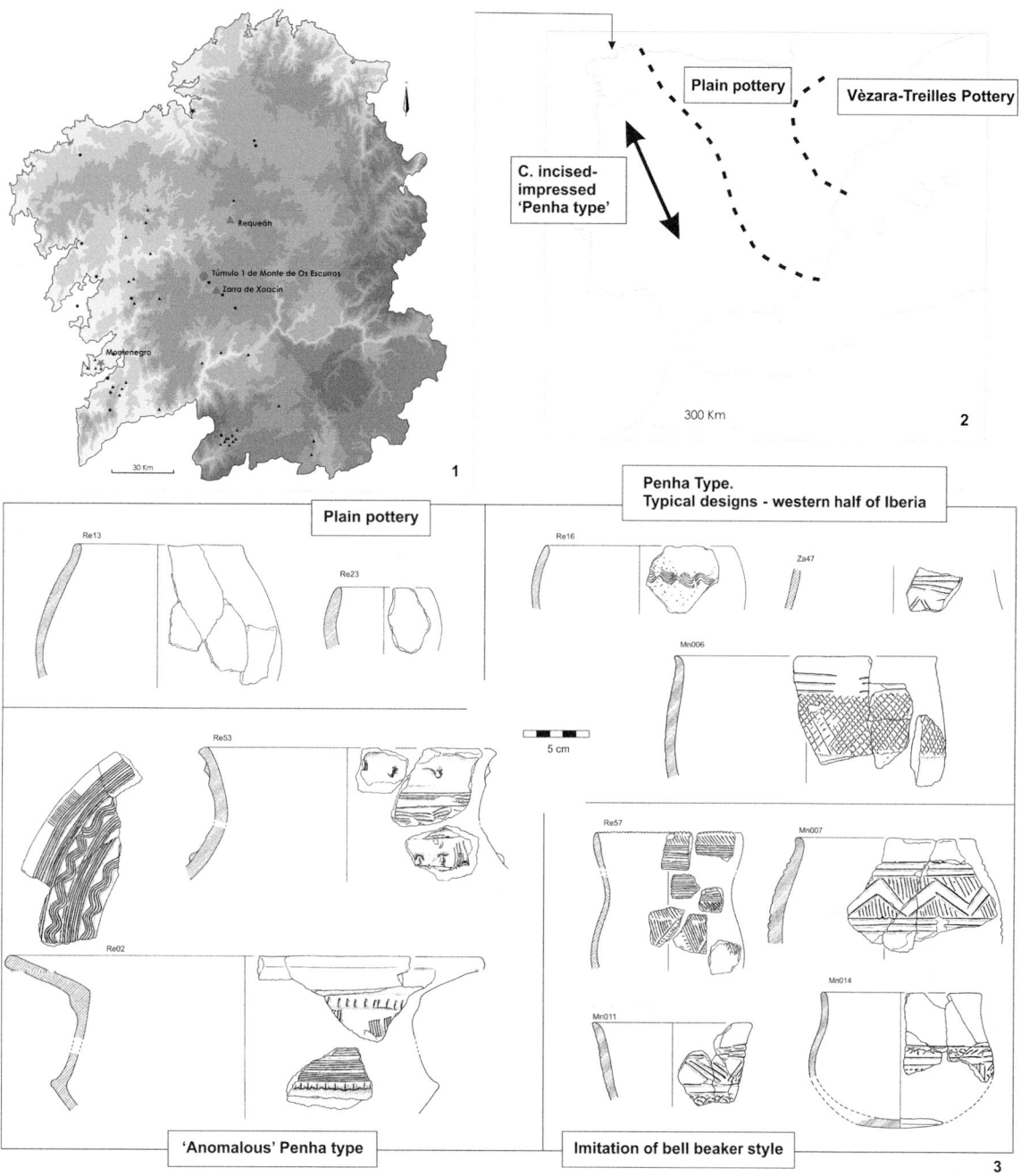

FIGURE 3. POTTERY ANALYSED FROM THE LATE NEOLITHIC. 1. MAP SHOWING THE DISTRIBUTION OF LATE NEOLITHIC SITES IN GALICIA (FROM PRIETO, 2004 AND 2009) IDENTIFYING THE SITES SELECTED FOR THE ANALYSIS. 2. MAP SHOWING THE DISTRIBUTION OF THE 3 POTTERY AREAS OF THE IBERIAN PENINSULA (FROM PRIETO, 2009). 3. SELECTION OF SOME OF THE VESSELS ANALYSED (ILLUSTRATION BY ANXO RODRÍGUEZ).

The archaeometric results have made it possible to identify 13 minerals (table 3a). The granitic composition (alkaline and calc-alkaline granites) is predominant as raw material in settlements, and to a lesser degree basic, amphibole and ultra-basic schists. In the tombs, the alkaline granitic composition is more limited in both vessels (although the number of samples is very small). In summary, all of the material from the Late Neolithic is granitic, except for a number of more varied compositions from the settlements (table 3b).

	Granitic				Metamorphic				Amphibolic and Ultrabasic				
	Q	FK	Pla	Mic	Clo	Epi	Cli	Ana	Anf	Ser	Cal	Px	Tal
Late Neolithic (domestic contexts, N=33)													
F%	100	88	97	48	33	6	–	3	73	12	–	–	–
Ab	33	24	29	5	9	12	–	3	13	5	–	–	–
Late Neolithic (funerary contexts, N=2)													
Te1b	24	8	40	2	5	–	–	–	10	–	–	–	3
Te4	27	33	23	1	–	–	–	–	–	–	–	–	–

	Alteration phyllosilicates						Iron oxides		
	Kao	Gib	Bay	Hal	Ver	Zeo	Hem	Goe	Ber
Late Neolithic (domestic contexts, N=33)									
F%	6	–	–	9	–	–	12	–	–
Ab	10	–	–	7	–	–	14	–	–
Late Neolithic (funerary contexts, N=2)									
Te1b	–	–	–	8	–	–	–	–	–
Te4	–	–	–	15	–	–	–	–	–

TABLE 3A. MINERALOGY OF THE POTTERY ANALYSED FROM THE LATE NEOLITHIC.

Context	In situ	Local	District	Regional
	0-1 km	< 7 km	7-50 km	50-200 km
Domestic	19	1	13	
Funerary		2		

TABLE 3B. MINIMUM DISTANCES FROM THE SITE TO THE MOST LIKELY SOURCES OF RAW MATERIAL FOR THE POTTERY FROM THE LATE NEOLITHIC.

The likely ceramic-source distances for the settlements vary between an *in situ* range and another district range (40%), while the local range is scarce. The funerary context is local, like some pieces from the previous period (although there are only 2 pieces).

A large number of different types of pieces have been identified from within the district range (7-50 km). We found pieces from all of the previously classified groups, noting a greater degree of mobility for undecorated pottery and Penha-type decorated pottery with Iberian features. It is possible that someone carried the product, which would explain frequent movement between the settlements in the district.

Out of the 36 pieces, 13 could have mobility at district level (36% of the selected pieces): 3 undecorated bowls, 1 undecorated pot, 1 anomalous Penha-type, 1 imitation bell beaker and 7 with regional Penha-type designs. Only one site is known in Portugal, Abrigo de Buraco da Pala (Sanches, 1997), with vessels that could be described as imitation bell beakers. These are the most individual types of vessels from the site, and no two are the same.

Early Bronze Age (2800/2600 -1600/1400 BC)

A total of 104 vessels from 20 sites have been selected, which are representative of all of the contexts (fig. 4.2). Overall we can differentiate between 3 main groups, which are characteristic at a pan-European level (described in detail in recent publications by Prieto, 2005, 2011 and 2013):

- Pottery with bell beaker decoration, in its standard and regional versions. In Galicia there are three varieties: standard, regional with comb and shell printing, and a grooved regional variety, and at least 14 varieties have been identified in the Iberian Peninsula.

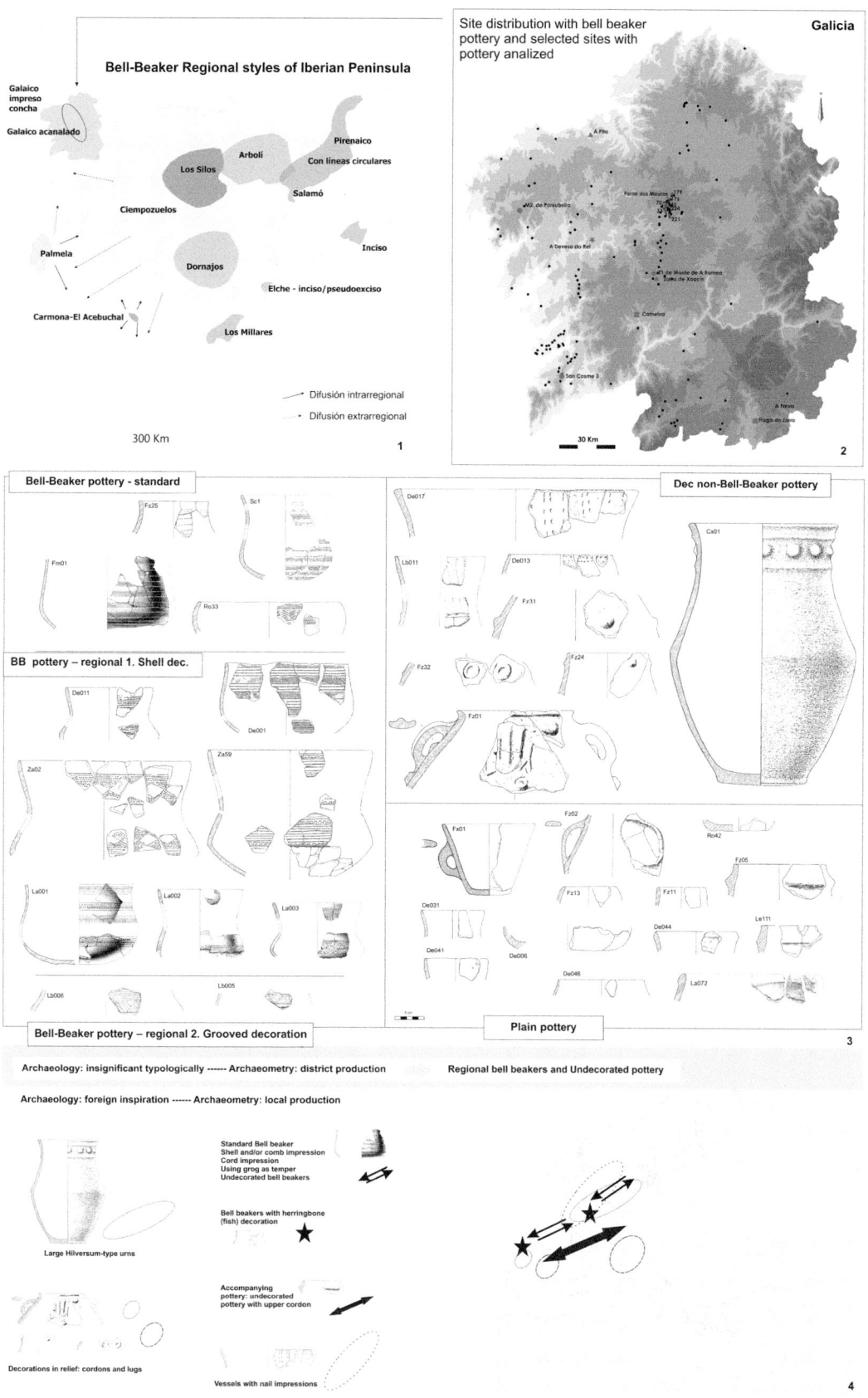

FIGURE 4. POTTERY ANALYSED FROM THE EARLY BRONZE AGE. 1. DISTRIBUTION OF REGIONAL BELL BEAKER STYLES IN THE IBERIAN PENINSULA. 2. MAP SHOWING THE DISTRIBUTION OF EARLY BRONZE AGE SITES IN GALICIA (FROM PRIETO, 2011) IDENTIFYING THE SITES CHOSEN FOR THE ANALYSIS. 3. SELECTION OF SOME OF THE VESSELS ANALYSED (ILLUSTRATION BY ANXO RODRÍGUEZ). 4. INTERPRETATION OF THE POSSIBLE LONG AND SHORT DISTANCE RELATIONSHIPS AT THIS TIME.

	Granitic				Metamorphic				Amphibolic and Ultrabasic				
	Q	FK	Pla	Mic	Clo	Epi	Cli	Ana	Anf	Ser	Cal	Px	Tal
Early Bronze Age (domestic contexts; N=58)													
F%	100	45	88	28	45	5	17	2	76	31	2	3	41
Ab	30	17	30	5	7	6	6	13	14	7	1	9	27
Early Bronze Age (ceremonial contexts, N=24)													
F%	100	79	92	83	21	–	–	13	38	–	8	–	25
Ab	40	23	16	9	34	–	–	5	22	–	1	–	3
Early Bronze Age (funerary contexts, N=43)													
F%	100	86	84	93	16	–	–	–	35	–	–	–	2
Ab	35	14	30	6	14	–	–	–	4	–	–	–	2

	Alteration phyllosilicates						Iron oxides		
	Kao	Gib	Bay	Hal	Ver	Zeo	Hem	Goe	Ber
Early Bronze Age (domestic contexts; N=58)									
F%	2	–	–	22	2	–	–	–	–
Ab	58	–	–	18	19	–	–	–	–
Early Bronze Age (ceremonial contexts, N=24)									
F%	13	–	–	21	–	–	4	–	–
Ab	9	–	–	10	–	–	6	–	–
Early Bronze Age (funerary contexts, N=43)									
F%	19	5	2	86	5	–	5	–	5
Ab	19	34	6	13	19	–	12	–	13

TABLE 4A. MINERALOGY OF THE POTTERY ANALYSED FROM THE EARLY BRONZE AGE.

Context	In situ	Local	District	Regional
	0-1 km	< 7 km	7-50 km	50-200 km
Domestic	10	29	19	
Ceremonial		16	8	
Funerary	4	7	31	1

TABLE 4B. MINIMUM DISTANCES FROM THE SITE TO MOST LIKELY SOURCES OF RAW MATERIAL FOR THE POTTERY FROM THE EARLY BRONZE AGE.

- Non-bell beaker decorated pottery: the smallest group in number, but with the widest variety of designs (braiding, lugs, printing using nails or punches).
- Undecorated pottery.

The archaeometric results have made it possible to identify 18 minerals (table 4a). This is the period when we identify a larger number of minerals, and it is likely that the number is larger because the sample studied is equally larger (table 4b).

Granitic compositions (alkaline and calc-alkaline) are predominant in all of the contexts, with greater diversity in the settlements and ceremonial sites, as ultra-basic compositions are documented in the first, and basic schists in the second. It is interesting to note the important influence of the geographical location of the sites in the compositions of the pottery.

Mainly local distances in settlements and ceremonial sites. There are no in situ productions in the ceremonial sites, while in the tombs there is a predominance of distances at district level (and even one at regional level), representing a significant percentage in all of the contexts.

Based on the available data, we believe that the types identified within a district distance range are once again varied in this period, representing 51% of the selected pieces. However, it is important to note that there are variations depending on the context.

In settlements (a total of 19) there are mainly undecorated bowls and cups (7), occasionally with braiding on their upper section (6) in the settlements (only 6 bell beakers from the regional variety decorated using shells and incisions were recorded in this group).

In the ceremonial sites (a total of 8) the undecorated vessels (4) have the same proportion as the bell beakers (3 in the regional variety and one in the herringbone variety).

Finally, in the funerary context (a total of 28), we find a majority of decorated non-bell beaker vessels from individual pit graves (10) and jar-type undecorated pottery, occasionally with braiding (9+1). There are only minimal remnants of bell beakers within a district range, including those from A Forxa, a necropolis of pit graves, or those from Parxubeira (3 standard type vessels with incrusted white clays).

In summary, there is greater mobility at district level for the undecorated pottery and to a much lesser extent for regional bell beakers from the settlements. We see greater mobility for undecorated and regional bell beakers in ceremonial sites. In funerary contexts with single pit-type graves, the vessels that were moved were decorated non-bell beakers and undecorated jars, while no mobility can be seen for the pan-European type pottery documented in all of the contexts, whose archaeometric results suggest that they were manufactured locally, although they do have clearly foreign influences based on a study of associated archaeological parallels, especially towards the north (France and England) and Central Europe (north Italy). The central tablelands of Spain and the north of Portugal can also be recognised as potential areas of contact (fig. 4.4).[2]

As a result, we can see that certain international types (bell beakers) are locally produced, while other vessels used on a more daily basis were produced at relatively long distances from the site. The greatest mobility of the objects is found in products for daily use or in regional versions of the European bell beaker.

The piece from a regional distance was probably used for spinning. It is small, and was something that was easy to carry as a personal item without it breaking.

Late Bronze Age (1600/1400 BC-900/700 BC)

We have selected 20 vessels from 7 sites in the region which are representative of all of the contexts (Prieto, 2005). Two main groups can be identified:

- Undecorated pottery is predominant for the period as a whole. A new shape appears, the wide horizontal rim, a specific shape to the NW Iberian Peninsula.
- The decorated pottery has simple designs made using deep or combed surface incisions. The printed technique is a new type of decoration from this stage, and so far has only been documented in wide horizontal rim pottery.
- Amongst this group we would draw attention to the presence of an exceptional vessel that was recently discovered, a *Cogotas* I-type vessel.

The archaeometric results have allowed us to identify 9 minerals (table 5a). In relation to raw materials, the domestic contexts are predominated by amphibole and granitic compositions (both

---

[2] These ideas are partly discussed in Prieto, 2011b; Prieto and Gil, 2011, and the parallels found in different regions have been taken from Besse, 2003; Fokkens, 2005; Gibson and Gill, 2013; Gibson and Snape, 2013; L'Helgouach, 1967, and Salanova, 2000.

	Granitic				Metamorphic				Amphibolic and Ultrabasic				
	Q	FK	Pla	Mic	Clo	Epi	Cli	Ana	Anf	Ser	Cal	Px	Tal
Late Bronze Age (domestic contexts, N=6)													
F%	100	100	100	100	–	–	–	–	67	–	–	–	–
Ab	37	22	20	5	–	–	–	–	11	–	–	–	–
Late Bronze Age (indefinite contexts, N=2)													
GU12A	8	16	72	–	–	–	–	–	–	–	–	–	–
GUL01	30	19	28	–	10	–	–	–	–	–	–	–	–
Late Bronze Age (funerary contexts; N=12)													
F%	100	92	92	83	25	–	–	8	33	–	–	–	17
Ab	32	27	14	5	33	–	–	3	22	–	–	–	6

	Alteration phyllosilicates						Iron oxides		
	Kao	Gib	Bay	Hal	Ver	Zeo	Hem	Goe	Ber
Late Bronze Age (domestic contexts, N=6)									
F%	50	–	–	17	–	–	–	–	–
Ab	13	–	–	15	–	–	–	–	–
Late Bronze Age (indefinite contexts, N=2)									
GU12A	–	–	–	5	–	–	–	–	–
GUL01	–	–	–	14	–	–	–	–	–
Late Bronze Age (funerary contexts; N=12)									
F%	42	–	–	42	8	–	–	–	–
Ab	7	–	–	12	6	–	–	–	–

TABLE 5A. MINERALOGY OF THE POTTERY ANALYSED FROM THE LATE BRONZE AGE.

Context	In situ	Local	District	Regional
	0-1 km	< 7 km	7-50 km	50-200 km
Domestic		2	4	
Indefinite	1		1	
Funerary		8	4	

TABLE 5B. MINIMUM DISTANCES FROM THE SITE TO MOST LIKELY SOURCES OF RAW MATERIAL FOR THE POTTERY FROM THE LATE BRONZE AGE.

alkaline and calc-alkaline). In the funerary contexts the compositions are the same as those in the settlements, although there is a predominance of granitic compositions and ultra-basic compositions are also found. In the unknown contexts there are only two pieces, one of which is granitic and the other schistose.

The domestic contexts received raw materials from within a local and district range, while the funerary contexts are predominated by local distance ranges. In the unknown context, one piece of pottery could have been made in situ, and the other corresponds to a district distance range (table 5b).

The types identified at district range only represent 40% of the selected pieces. They demonstrate mobility over longer distances, with smaller, mainly undecorated vessels from settlements. It can be seen that the vessels associated with graves with what is probably a district distance range are found in the re-use of collective megalithic tombs.

In this period we can highlight two types of vessels:

The first type, the *wide horizontal rim vessels* (WHR), with a regional personality whose raw material comes from a local or *in situ* distance range, coherent with the territorial peculiarity of this type of

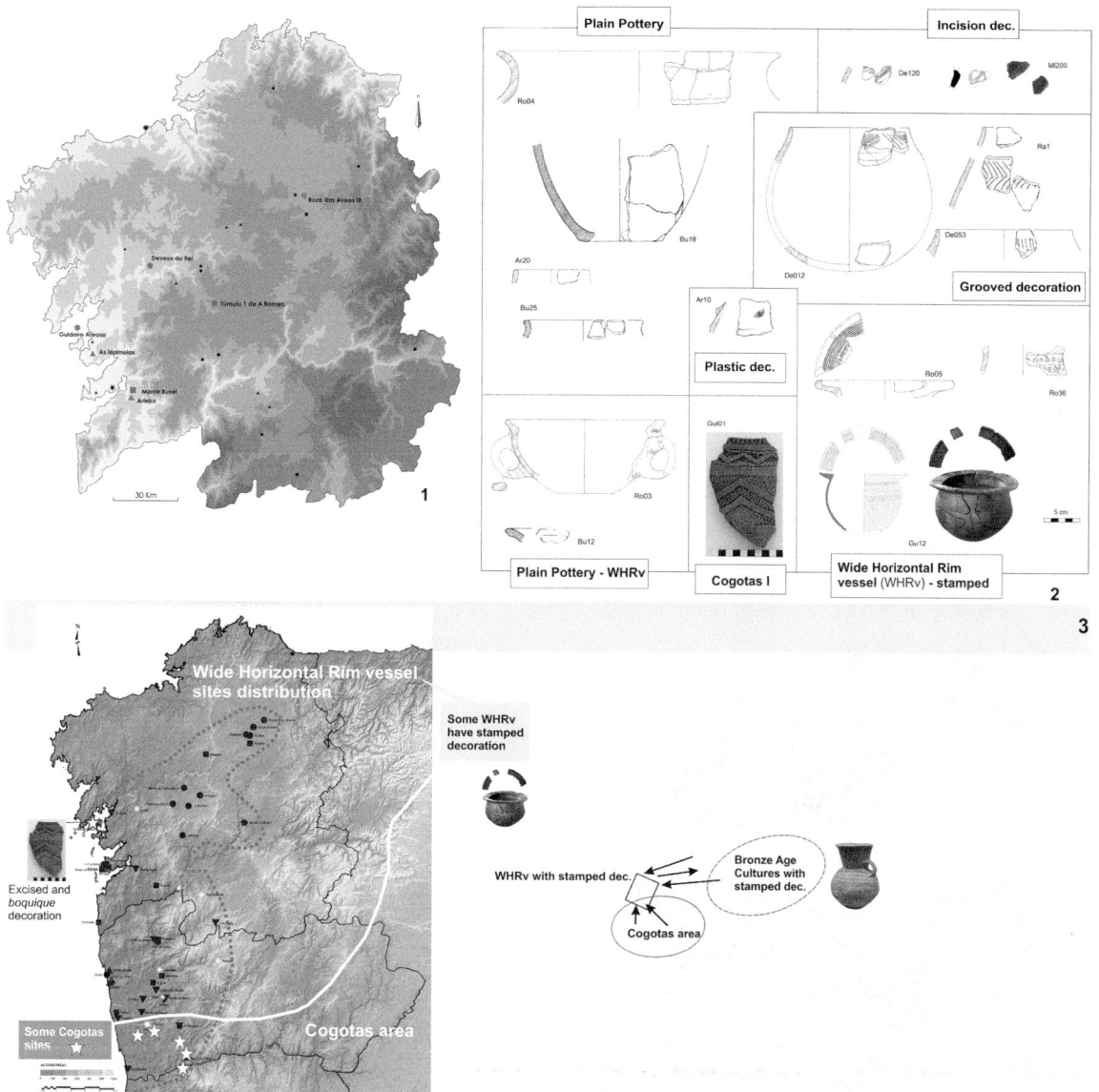

FIGURE 5. POTTERY ANALYSED FROM THE LATE BRONZE AGE. 1. MAP SHOWING THE DISTRIBUTION OF LATE BRONZE AGE SITES (FROM PRIETO, 2005), SHOWING THE SITES SELECTED FOR THE ANALYSIS. 2. SELECTION OF SOME OF THE VESSELS THAT WERE ANALYSED (ILLUSTRATION BY ANXO RODRÍGUEZ, PHOTO BY PILAR PRIETO). 3. DISTRIBUTION OF SITES WITH WHR POTTERY IN THE NW IBERIAN PENINSULA (FROM NONAT ET AL., 2015) AND INTERPRETATION OF THE POSSIBLE RELATIONSHIPS WITH OTHER PARTS OF EUROPE THROUGH THE STAMPED DECORATION ON THE WHR POTTERY.

pottery (Nonat *et al.*, 2015). The stamped decoration found on a small number of vessels points towards the possibility that the person who decorated them may have once again come from France (Gomez de Soto 1995), or learned the technique from a potter from this area.

The second, the *Cogotas I-type* fragment, a unique piece in the region (with excised and *bouquique* decoration), which the analytical results identify as a seemingly foreign product manufactured at a district distance range. This vessel comes from a Cogotas I-type tradition, characteristic of the central tablelands and other parts of the peninsula (Blanco González, 2011). In its phase of expansion outside of the central tablelands, the chronology of this pottery developed from 1500 BC onwards (Galán Saulnier, 1998; López-Romero *et al.*, 2015). The closest site with Cogotas-type pottery is further away than the analyses are capable of evaluating, in the settlements of A Sola (Bettencourt, 1991-2) or Bouça do Frade (Jorge, 1988), which are less than 200 km to the south, on the frontier of the region with wide horizontal rim pottery. The manufacturing processes used for this type of pottery do not coincide with Galician pottery produced in the Late Bronze Age, which means that a person could have exchanged it until it reached the site, or could have placed it directly in the place where it was found.

5. Final comments

Although we need to exercise caution in terms of the results, as a small number of pieces were selected in relation to the amount of material present in the region, we can put forward a number of interesting and important ideas:

1. We have broken away from the binomial idea of local-foreign, defining more probable distance ranges.
   - We would draw attention the district range (7-50 km) as an important value in the definition of mechanisms of circulation, something that is very frequent/usual in any period.

2. Greater or lesser mobility over time.
   - Although with the Bell Beaker it would seem that more complex and varied mechanisms of circulation were set underway in the region.

3. There is a greater mobility of products than expected.
   - The most special vessels were not those that were moved the most.
   - The vessels that were moved the most were those used on a daily basis or in regional style.

4. There was probably widespread movement of large numbers of people over variable distances, in complex and constant networks of communication.

With regard to the methodology used, we would highlight the following. Firstly, combining archaeometry and archaeology offers great potential for research.

Secondly, using archaeometric criteria, pieces were selected that would otherwise never have been considered if the selection had only be based on archaeological criteria, as we have seen that the materials that have greater mobility from an archaeometric perspective do not have this mobility based on their typology.

Finally, archaeometry provides information on the types of mobility at district level, and although it does not provide a direct answer, it does help to correct preconceived ideas through archaeology, and to ask new questions.

As a future outlook, we would highlight the fact that although the sample is not distributed evenly over the different periods, it is possible to distinguish certain diachronic trends. The next step will be to include the rest of the vessels that have already been studied using an archaeological methodology, even though no analyses are available, in order to complete the information which is highly significant in its own regard with the sample shown in this study, and of course to continue with the analytical programme, adding new vessels to the sample.

Acknowledgements

This study has been funded through the project *Isótopos de Pb y Sr en cerámicas arqueológicas de Galicia: estudio de la procedencia y el acceso a las materias primas (EM2012/054, 2012-PG217)*, 2012-2015, as a part of the Call for Funding for Research Projects and Emerging Researchers of the Galician RandDandI Plan, from the Regional Government of Galicia (Xunta de Galicia).

References

Alday Ruiz, A.; Carvalho, A.-F.; Cerrillo Cuenca, E.; Gonzálezcordero, A.; Juez Aparicio, L.; Moral del Hoyo, S.; Ortega Martínez, A.-I., 2009 – Reflejos del neolítico Ibérico. La cerámica boquique: caracteres, cronología y contexto. Milán: Edar Arqueología y Patrimonio. 179 p.

Besse, M., 2003 – L'Europe du 3ème millénaire avant notre ère: les céramiques communes au Campaniforme. Lausanne: Cahiers d'Archéologie Romande 94. 223 p.

Bettencourt, A.-M.-S., 1991-1992 – O povoado da Sola, Braga: noticia preliminar das escavações de 1991-92. Cadernos de Arqueologia. Braga. Série II: 8-9, pp. 97-118.

Blanco González, A., 2011 – Práctica social, memoria y ritual en Cogotas I: esbozo teórico para un enfoque renovado. Trabajos de Prehistoria. Madrid. 68: 1, pp. 123-146.

Cassen, S.; Boujot, C. H.; Dominguez Bella, S.; Guiavarc'h, M.; Le Pennec, C. H.; Prieto Martínez, M.-P.; Querré, G.; Santrot, M.-H.; Vigier, E., 2012 – Chapitre 16. Dépôts bretons, tumulus carnacéens et circulations à longue distance. In: JADE. Grandes haches alpines du néolithique européen Ve au IVe millénaires av. J.-C., deuxième partie: les haches en jades; de l'Italie a l'Atlantique, tome 2. *Les cahiers de la MSHE Ledoux*. pp. 918-995.

Cobas Fernández, I.; Prieto Martínez, M.-P., 2003 – The technological chain as a methodological and theoretical tool from archaeology. In: Le Secretariat Du Congrès (Ed), Section 1- Theory and Methods. General Sessions And Posters. Oxford: B.A.R. pp. 1-8. (BAR International Series, 1145).

Fokkens, H., 2005 – Le début de l'Âge du Bronze aux Pays-Bas et l'Horizon Hilversum ancien. In: Bourgeois, J.; Talon, M., eds.- L'Âge du Bronze du Nord de la France dans son contexte européen. Paris: Éditions du Comité des Travaux Historiques et Scientifiques. pp. 11-33.

Galan Saulnier, C., 1998 – Sobre la cronología de Cogotas I. Cuadernos de Prehistoria y Arqueología de la Universidad Autónoma de Madrid. Madrid. 25: 1, pp. 201-243.

Gibson, A.; Gill, D., 2013 – Chapter 20: Beaker occupation at Cavenham Quarry, Suffolk. In: Prieto Martínez, M. P.; Salanova, L., coords.- Current researches on Bell Beakers. Proceedings of the 15th International Bell Beaker Conference: From Atlantic to Ural. 5th – 9th May 2011 Poio Pontevedra, Galicia, Spain Santiago de Compostela: Copynino-Centro de Impresión Digital. pp. 251-264.

Gibson, A.; Snape, N., 2013 – Chapter 12: Two beaker vessels from Maxey Quarry, Cambridgeshire. In: Prieto Martínez, M. P.; Salanova, L., coords. – Current researches on Bell Beakers. Proceedings of the 15th International Bell Beaker Conference: From Atlantic to Ural. 5th – 9th May 2011 Poio Pontevedra, Galicia, Spain Santiago de Compostela: Copynino-Centro de Impresión Digital. pp. 129-130.

Gomez de Soto, J., 1995 – Le Bronze Moyen en Occident. La culture des Duffaits et la Civilisation des Tumulus. Paris: Picard (L'âge du Bronze en France 5). 375 p.

Jorge, S. O., 1986 – Povoados da Pré-história recente da região de Chaves-V^a. P^a. de Aguiar. Porto: Cámara Municipal de Chaves. 3 vols. 1131 p.

Jorge, S. O., 1988 – O povoado da Bouça do Frade (Baião) no quadro do Bronze Final do Norte de Portugal. Porto: Grupo de Estudos Arqueológicos do Porto (Monografías Arqueológicas 2). 124 p.

Kaal, J.; Lantes-Suárez, O.; Martínez Cortizas, A.; Prieto Lamas, B.; Prieto Martínez, M.-P., 2013 – How useful is pyrolysis-GC-MS for the assessment of molecular properties of organic matter in archaeological pottery matrix? An exploratory case study from North-West Spain. Archaeometry. Oxford. 56: 1, p. 187: 207 (DOI: 10.1111/arcm.12057).

L'Helgouach, J., 1967 – Le monument mégalithique à entrée latérale de Crec'h Quillé en Saint-Quay-Perros (Côtes-du-Nord). Bulletin de la Société Préhistorique Française. Paris. 64, pp. 659-698.

Lantes-Suárez, O.; Prieto Martínez, M. P. and Martínez Cortizas, A., 2010 – Caracterización de la pasta blanca incrustada en decoraciones de campaniformes gallegos: Indagando sobre su procedencia. In: Saiz Carrasco, M^a.; López Romero, R.; Cano Díaz-Tendero, Mª A.; Calvo García, J. C. eds.- VIII Congreso Ibérico de Arqueometría. ACTAS (Teruel, 19-21 Octubre 2009). Teruel: Ed. Seminario de Arqueología y Etnología Turolense. pp. 87-100.

Lantes-Suárez, O.; Prieto Martínez, M-P.; Martínez Cortizas, A., 2011 – Aplicación de la Microscopía Electrónica de Barrido al estudio de los acabados de cerámica antigua de Galicia. Gallaecia. Santiago de Compostela. 30, pp. 117-125.

López-Romero, E.; Gümil-Fariña, A.; Mañana-Borrazás, P.; Otero Vilariño, C.; Prieto Martínez, M.-P.; Rey García, J.-M.; Vilaseco Vázquez, X.-I., 2015 forthcoming – El conjunto arqueológico de Guidoiro Areoso (Ría de Arousa, Pontevedra): síntesis y perspectivas actuales. Trabajos de Prehistoria. Madrid.

Martínez Cortizas, A.; Lantes-Suárez, O.; Prieto Martínez, M.-P., 2010 – Análisis arqueométrico de la cerámica de contextos campaniformes del Área Ulla-Deza. In: Prieto Martínez, M. P.; Criado-Boado, F. coords.- Reconstruyendo la historia de la comarca del Ulla-Deza (Galicia, España). Escenarios arqueológicos del pasado. (TAPA 41). Madrid. CSIC. pp. 135-144.

Martínez Cortizas, A.; Prieto Lamas, B.; Lantes-Suárez, O.; Prieto Martínez, M.-P., 2008 – 'Análisis mineralógico, elemental y cromático de cerámica prehistórica del área Ulla-Deza (Noroeste de la Península Ibérica)'. In: Rovira Llorens, S.; García Heras, M.; Gener Moret, M.; Montero Ruiz, I., eds.- Actas del VII Congreso Ibérico de Arqueometría (Madrid, 8-10 octubre 2007). pp. 250-264. (http://www.ih.csic.es/congreso_iberico/index.PDF).

Martínez Cortizas, A.; Lantes-Suárez, O.; Prieto Martínez, M.-P., 2011 – Capítulo 33. Cerámica campaniforme del NW de la Península Ibérica. Indagando en sus materias primas, elecciones tecnológicas y procedencia. In: Prieto Martínez, M. P.; Salanova, L., coords.- Las Comunidades Campaniformes en Galicia. Cambios sociales en el III y II milenios BC en el NW de la Península Ibérica. Pontevedra. Diputación de Pontevedra. pp. 309-331.

Nonat, L.; Vázquez Liz, P.; Prieto Martínez, M.-P., 2015 – El vaso de largo bordo horizontal: un trazador cultural del noroeste de la península ibérica en el II milenio BC. Oxford. Archaeopress, British Archaeological Reports International Series 2699. 185 p.

Prieto Lamas, B.; Lantes-Suárez, O.; Prieto Martínez, M.-P.; Martínez Cortizas, A., 2011 – Capítulo 34: Aspectos cromáticos de la cerámica campaniforme de Galicia. In: Prieto Martínez, M. P.; Salanova, L., coords.- Las Comunidades Campaniformes en Galicia. Cambios sociales en el III y II milenios BC en el NW de la Península Ibérica. Pontevedra. Diputación de Pontevedra. pp. 333-343.

Prieto Martínez, M.-P., 2004 – Ceramic style in Neolithic societies in Galicia (NW Iberian Peninsula). Similarities and differences in patterns or formal regularity. In: Section 9- The Neolithic in the Near East and Europe/ Section 10 The Copper Age in the Near East And Europe. General Sessions and Posters. Oxford: B.A.R. pp. 109-117. (BAR International Series, 1303).

Prieto Martínez, M.-P., 2005 – Ceramic style in Bronze Age societies in Galicia (NW Iberian Peninsula). Similarities and differences in patterns or formal regularity. In: XIVth Congress of the U.I.S.P.P. (2-8 Septiembre 2001, Liège). Section 11- L'âge Du Bronze En Europe et En Mediterranee/ The Bronze Age in Europe and the Mediterranean. General Sessions and Posters. Oxford: B.A.R. pp. 99-107. (BAR International Series, 1337).

Prieto Martínez, M.-P., 2009 – Chapter V. From Galicia to the Iberian Peninsula: Neolithic ceramics and traditions. In: Dragos, G., ed.- Early farmers, Late Foragers and Ceramic traditions. On the beginning of pottery in Europe. Cambridge.Cambridge Scholars Press. pp. 116 -149.

Prieto Martínez, M.-P., 2010 – 'La cerámica de O Regueiriño (Moaña, Pontevedra). Nueva luz sobre el neolítico en Galicia'. Gallaecia. Santiago de Compostela. 29, pp. 63-82.

Prieto Martínez, M.-P., 2011a – Capítulo 35. Alfarería de las comunidades campaniformes en Galicia: contextos, cronologías y estilo. In: Prieto Martínez, M. P.; Salanova, L., coords.- Las Comunidades Campaniformes en Galicia. Cambios sociales en el III y II milenios BC en el NW de la Península Ibérica. Pontevedra. Diputación de Pontevedra. pp. 345-362.

Prieto Martínez, M.-P., 2011b – Capítulo 15. La urna de Cameixa ¿una influencia del Horizonte Hilversum?. In: Prieto Martínez, M. P.; Salanova, L., coords.- Las Comunidades Campaniformes en Galicia. Cambios sociales en el III y II milenios BC en el NW de la Península Ibérica. Pontevedra. Diputación de Pontevedra. pp. 127-131.

Prieto Martínez, M.-P., 2012 – Perceiving changes in the third millennium cal. BC in Europe through pottery: Galicia, Brittany and Denmark as an example. In: Prescott, C. H.; Glørstad, H., eds.-

Becoming European. Transformation of third millennium Northern and Western Europe. Oxford. Oxbow Books. pp. 30-47.

Prieto Martínez, M.-P., 2013 – Chapter 19: Unity and circulation: what underlines the homogeneity of Galicia bell beaker ceramic style? In: Prieto Martínez, M. P.; Salanova, L., coords.- Current researches on Bell Beakers. Proceedings of the 15th International Bell Beaker Conference: From Atlantic to Ural. 5th – 9th May 2011 Poio Pontevedra, Galicia, Spain. Santiago de Compostela. Copynino-Centro de Impresión Digital. pp. 209-249.

Prieto Martínez, M.-P.; Gil Agra, D., 2011 – Capítulo 17. Fraga do Zorro: fosas y cacharros. Innovaciones en la alfarería de la necrópolis. In: Prieto Martínez, M. P.; Salanova, L., coords.- Las Comunidades Campaniformes en Galicia. Cambios sociales en el III y II milenios BC en el NW de la Península Ibérica. Pontevedra. Diputación de Pontevedra. pp. 139-147.

Prieto Martínez, M.-P.; Lantes-Suárez, O.; Martínez Cortizas, A., 2008 – O Campaniforme Cordado de Forno dos Mouros (Toques, A Coruña). Cuaderno de Estudios Gallegos. Santiago de Compostela. LV: 121, pp. 31-51.

Prieto Martínez, M.-P.; Lantes-Suárez, O.; Vázquez-Liz, P.; Martínez Cortizas, A., 2010 – La cerámica de dos túmulos de Roza das Aveas (Outeiro de Rei, Lugo): Un estudio diacrónico del estilo y la composición. BSAA Arqueología. Valladolid. LXXVI, pp. 27-62.

Prieto Martínez, M.-P.; Lantes-Suárez, O.; Martínez Cortizas, A., 2009b – Dos enterramientos de la Edad del Bronce en la provincia de Ourense. Revista Aquae Flaviae. Chaves. 41, pp. 93-105.

Prieto Martínez, M. P.; Mañana Borrazás, P.; Costa Casais, M.; Criado-Boado, F.; López Sáez, J. A.; Carrión Marco, Y.; Martínez Cortizas, A., 2012 – El Neolítico en Galicia. In: M. Rojo, A. M.; Garrido, R.; García, I., coords.- El Neolítico en la Península Ibérica y su contexto europeo. Cátedra. Colección Historia. Serie Mayor. pp. 30-47.

Prieto Martínez, M.-P.; Martínez Cortizas, A.; Lantes-Suárez, O.; Gil Agra, D., 2009a – Estudio de la cerámica del yacimiento de fosas de Fraga do Zorro. Revista Aquae Flaviae. Chaves. 41, pp. 107-121.

Prieto Martínez, M.-P.; Salanova, L., 2009 – Coquilles et Campaniforme en Galice et en Bretagne: mécanismes de circulation et stratégies identitaires. Bulletin de la Société Préhistorique Française. Paris. 106: 1, pp. 73-93.

Prieto Martínez, M.-P.; Vazquez Collado, S., 2011 – Capítulo 29. Campaniformes singulares ¿imitación u ocultación [diferenciación] de la identidad? In: Prieto Martínez, M. P.; Salanova, L., coords.- Las Comunidades Campaniformes en Galicia. Cambios sociales en el III y II milenios BC en el NW de la Península Ibérica. Pontevedra. Diputación de Pontevedra. pp. 267-274.

Rojo, M.-A.; Garrido, R.; García, I. (coords.), 2012 – El Neolítico en la Península Ibérica y su contexto europeo. Madrid: Cátedra, Colección Historia, Serie Mayor. 580 p.

Salanova, L., 2000 – La question du campaniforme en France et dans les Îles anglo-normandes: productions, chronologie et rôles d'un standard céramique. Paris: Coédition Société Préhistorique Française et Comité des Travaux Historiques et Scientifiques. 392 p.

Salanova, L.; Prieto Martínez, M.-P.; Clop, X.; Convertini, F.; Lantes-Suárez, O.; Martínez Cortizas, A., 2015 – What are large-scale archaeometric programmes for? Bell Beaker pottery and societies from the 3rd millennium BC in Western Europe. Archaeometry. Oxford.

Sanches, M. J., 1997 – A Pré-História Recente de Trás-os-Montes e Alto Douro. Porto: Sociedade Portuguesa de Antropologia e Etnologia. 2 vols. 331 p.

Types and gesture. The jewellery of the Copper age in the Alps in a techno-typological study

Stefano Viola[1], Maria Adelaide Bernabo' Brea[2], Dino Delcaro[3],
Federica Gonzato[4], Cristina Longhi[5], Giorgio Gaj[3], Roberto Macellari[6],
Luciano Salzani[4], Alessandra Serges[7], Iames Tirabassi[6] and Marie Besse[1]

[1] Laboratoire d'archéologie préhistorique et anthropologie – Département F.-A. Forel des sciences de l'environnement et de l'eau, Université de Genève
[2] Soprintendenza per i Beni archeologici dell'Emilia-Romagna
[3] Centro di Archeologia Sperimentale Torino
[4] Soprintendenza per i Beni archeologici Veneto
[5] Soprintendenza per i Beni archeologici della Lombardia
[6] Musei Civici di Reggio Emilia
[7] Museo Nazionale Preistorico Etnografico 'Luigi Pigorini' di Roma

Abstract

This contribution aims to compare jewellery artefacts from some northern Italy archaeological sites, dated to different periods: the Copper age and Early Bronze age. Through a techno-typological and functional study that takes into account several morphometric, morphological and specific trace parameters (indicators of anthropic and/or wear activity), the methods, techniques and tools are reconstructed and compared. On one hand, with the typological analysis, jewellery has been looked at as a cultural marker allowing to gather information (raw material, forms, and measures) on different aspects of past life, such as style, territories, and traditions. On the other hand, with the technological analysis, interpretative hypotheses are proposed based on the comparison between production traces and experimental data, in order to reconstruct (in part or completely) manufacture procedures and fabrication techniques. Finally, a functional analysis enabled to distinguish wear traces from technological traces and to recognize if the object has been used or not.

Key words: *Copper age, Functional analysis, Italy, Jewellery, Technology*

Résumé

Cette contribution vise à comparer les objets de parure provenant de certains sites archéologiques de l'Italie du nord, datés à différentes périodes: âge du Cuivre et Bronze ancien.

Grâce à une étude techno-typologique et fonctionnelle qui tient compte de plusieurs paramètres morphométriques, morphologiques et des traces spécifiques (indicateurs de l'activité de fabrication et/ou de l'utilisation) les méthodes, les techniques et les outils sont reconstruits et comparés. D'une part, la parure a été considérée, du point de vue typologique, comme un marqueur culturel permettant d'obtenir des informations (matière première, formes, mesures) sur les différents aspects de la vie du passé, tels que les styles, les territoires et les traditions. D'autre part, d'un point de vue technologique, des hypothèses interprétatives sont proposées sur la base de la comparaison entre les traces de fabrication et les données expérimentales dans le but de reconstruire (en partie ou complètement) les procédures de fabrication et les techniques employées. Enfin, une analyse fonctionnelle a permis de distinguer les traces d'usure des traces technologiques, et de reconnaitre l'utilisation ou non de l'objet.

Mots clés: *Age du Cuivre, Analyse fonctionnelle, Italie, Technologie, Parure*

Introduction

This paper aims to show a technological study of some elements of ornaments in stone. These artefacts belong to northern Italy inhumation burials and are dated from Copper to Early Bronze Age (see table 1). This study is part of a doctoral thesis at the University of Geneva, under the direction of prof. M. Besse (Viola 2016).

Site	Grave	Period	Number of objects	Bibliography
Remedello, BS	37	Copper	13	Cornaggia Castiglioni 1971, p. 61
Scalucce, VR	2	Copper	124	Valzolgher, Lincetto 2001-2002, fig. 10
Guidorossi, PR	3	Copper	17	Bernabo' Brea, Miari 2013, fig. 7
Arano, VR	49	Early Bronze	12	de Marinis, Valzolgher 2013, fig. 8

TABLE 1. THE LIST OF SITES.

The sites selected for the study offer a reliable chronological attribution and a set of homogeneous and numerically limited materials (166 beads in total), which makes this a suitable sample for techno-functional study. The materials of the first two sites, i.e. the grave 37 of the necropolis of Remedello Sotto and the grave 2 of the necropolis of Scalucce di Molina, although coming from old excavations (nineteenth century) are certainly attributable to the Copper age (Farioli *et al.*, 2015; Valzolgher, Lincetto 2001-2002). Both are funerary contexts of great importance and, in particular, the site of Remedello is a reference for the chrono-cultural articulation of the period, since it is the largest necropolis in northern Italy (de Marinis 2013). The materials of the other two sites come from very recent excavations (last ten years) and are chrono-stratigraphically reliable. While the beads of the tomb of via Guidorossi in Parma certainly date back to the Copper age, before the Bell Beaker Culture, those of the necropolis of Arano come from the most vast necropolis of Early Bronze age of the northern of Italy (Bernabo' Brea, Miari 2013; de Marinis, Valzolgher 2013).

Methods

The techno-functional study aims to apply an analytic investigation procedure that is effective to the identification of drilling marks – a very delicate stage with different variables – and at least of some aspects of the final phase of the manufacture through the use of a 'portable instrumentation.'

The interpretive work is based on the identification of criteria distinguishing drilling as well as surface modification techniques (Viola 2016). We tried to identify and record:

- Raw materials (determinations by the study of macroscopic characteristics).
- Main traces of manufacture.
- Main traces of use-wear.
- All possible tools involved in the production through the identification of some mark of wear and any associated archaeological objects (points, plate, etc.).

The techno-functional analysis was carried out through several stages:

1. Observation of objects (devices: stero microscope Seben Incognita, magnification: 620x, 40x, 50x; ocular x10, x20; lens: x2, x4; USB digital microscope Dino-Lite PRO digital microscope AM-413-T, magnification: 20x-220x, resolution: 1280x1024 pixels; Megapixels: 1.3 MP);
2. Photographic documentation of traces (device: USB digital microscope);
3. Comparison with traces of experimental tests. These tests are essential to create an experimental *corpus* of reference using raw materials, techniques and instruments chronologically and culturally compatible.

In addition to the morphological comparison of techno-functional use-wear marks, we considered some morphometric parameters (diameter, length, width, thickness, etc.). We also proposed some hypothesis based on the state of the surfaces according to macroscopic and microscopic features.

Methodologically, the technological study is based on the recognition of the marks (macro and micro marks) present on different surfaces. The overlap of the marks allows to reconstruct the manufacturing sequences. Each sequence has been described in terms of techniques, methods and tools used. Assuming that traces produced by different techniques and tools can be identified in archaeological materials, the study is based on an experimental framework around two different aspects: perforations and surface treatment. Both of them take into account several variables (see table 2). In total, about 200 tests were carried out on stones of different degrees of hardness.

Aspect	Variable	Description
P.- S.T.	Worked raw material	soapstone, limestone, calcite, marble
S.T.	Movement	transversal, longitudinal, etc…
P.- S.T.	Mechanism and instrument	hand drill, bow drill, pump drill, brace drill, polisher, grooved stone
P.- S.T.	Instrument raw material	bone, metal, stone, thorn, wood
P.- S.T.	Instrument morphology	axial, dejete, etc…
P.- S.T.	Abrasive (yes, no)	river sand, quartzite
P.- S.T.	Lubricant (yes, no)	water
P.- S.T.	Posture	sitting, standing

TABLE 2. THE EXPERIMENTAL TESTS: ASPECTS, VARIABLES AND DESCRIPTIONS.

Results: experimental framework

As an illustration of the results achieved in the experimental framework, we suggest the exemple of the tools used in the manufacturing processes. We show data obtained in a highly specific field concerning the wear during drilling phase. In the case of the hole-making tools, experiments were performed on lithic and metal tips (see table 3 and figure 1).

N°drill bit	Raw material	Worked raw material	Time	Motion
17	silex	soapstone	2'	bow drill
4	silex	marble	33'	hand drill/bow drill
		soapstone	4'	hand drill
			TOT. 37'	
12	bronze	soapstone	20'25'	bow drill

TABLE 3. THE PERFORATION TESTS.

Typologically, the lithic borers are axial, symmetrical and with a well-defined retouched bit (P4, P17) while the metal borers are axial, symmetrical and with a faceted bit (P12).

By comparing the tips before and after use, some specific features emerged according to some previous works (Bains 2013; Calley, Grace 1988; Chelidonio 1988; Coskunsu 2008; Gurova *et al.*, 2014, 2013).

In the case of non-perishable tips, these features can be divided in two types:

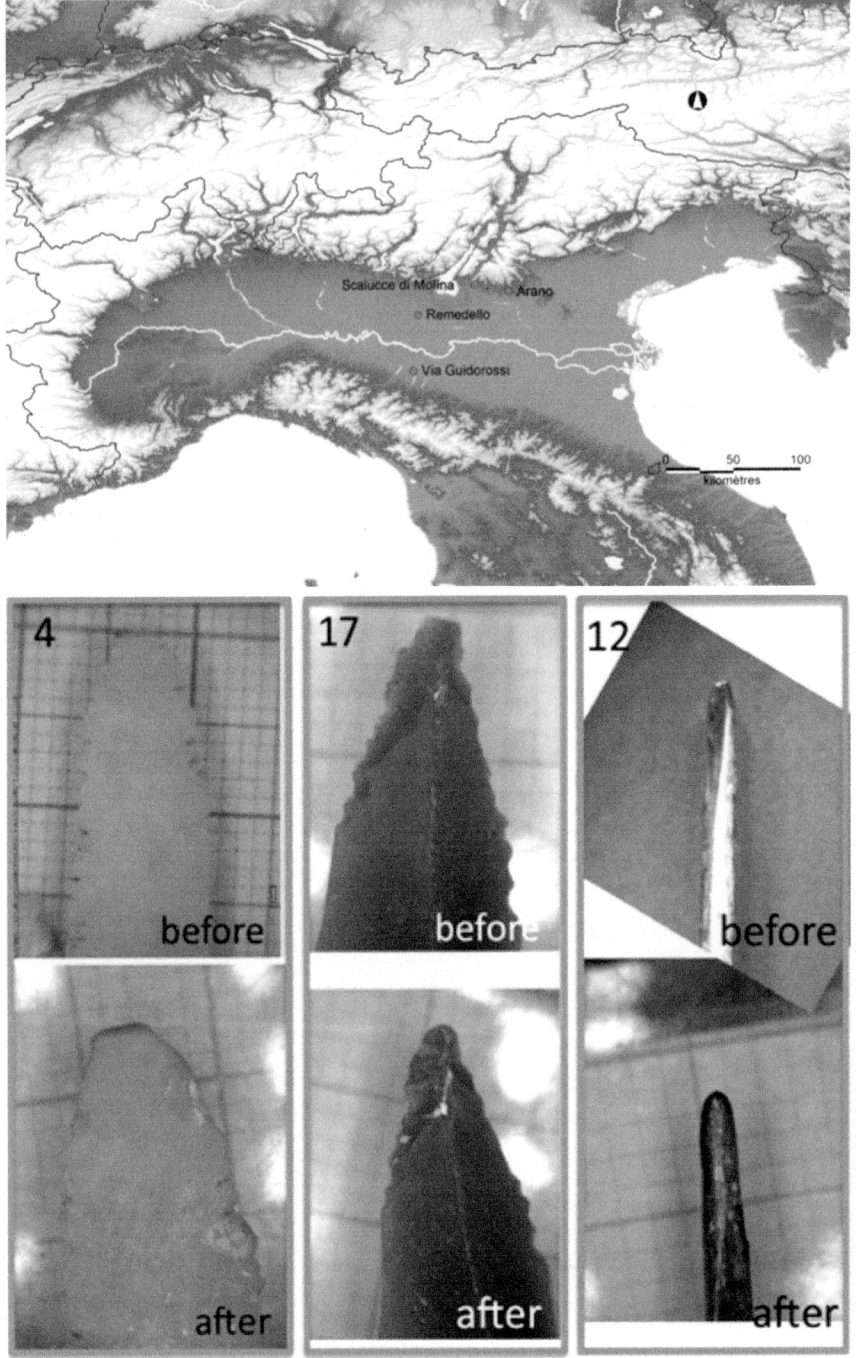

FIGURE 1. UP: LOCALIZATION OF THE SITES STUDIED; DOWN: HOLE MAKING TOLLS BEFORE AND AFTER USE.

morphological features:

- the tool shape must be developed along its axis
- the proximal end must be suitable to be handled in the rod

microscopic features – (either or any combinations of these features)

- Concentric circle on cross-section of drill bit
- Rounded tip
- Intensive surface polish
- Edge damage – only on lithic tips

As a practical application in the field of archaeology, these characteristics can help us to recognize the tips from archaeological context. At the same time, they could suggest the type used for perforations.

Results: techno-functional study

The techno-functional study allowed us to identify both the working and use-wear traces. In particular, some objects have clear working traces attributable to specific stages of the manufacturing sequence. This allows us to offer some reference operating ranges for each studied site. Below, we illustrate the results obtained in a schematic form through the proposal of a *chaîne opératoire* and some use-wear interpretations.

Copper age: disc and cylindrical beads, medium size (figure 2)

Archaeological context: grave 37-Remedello Sotto

Raw material: 11 carbonate, 1 soapstone

Morphometry: objects fairly standardized (11 disc, 1 cylindrical)

Manufacture sequences:

1. Roughing. Individual by abrasion (irregular outline of the beads) and mass calibration (side straight).

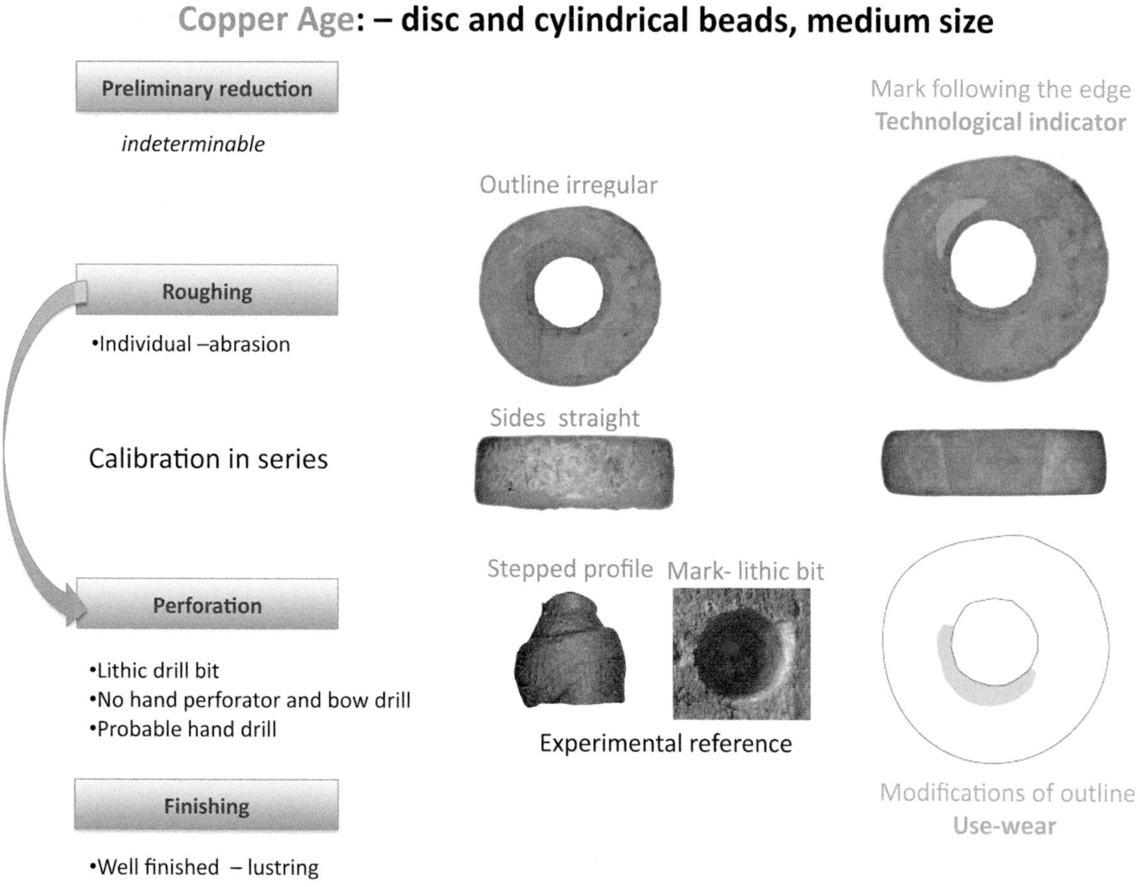

FIGURE 2. MANUFACTURE SEQUENCE OF COPPER AGE, DISC AND CYLINDRICAL BEADS-MEDIUM SIZE.

2. Perforation. Lithic drill bits (conical and stepped profile of the hole, mark following the edge); several drill bits (diameters not standardized) and methods (unipolar and bipolar); not reamed; no use of hand perforator and bow drill, probably use of hand drill (circular outline of perforations)
3. Finishing. Well finished by lustring

Use wear: normally beads show use wear (modifications of the perforation's outline).

Copper age: disc and cylindrical beads, small size (figure 3)

Archaeological context: grave 2-Scalucce di Molina

Raw material: 124 carbonate

Morphometry: objects fairly standardized (68 disc, 54 cylindrical)

Manufacture sequences:

1. Roughing. In series by abrasion (longitudinal striations, sides and edges straight). Probably from rod (difficulty in centring the holes).
2. Perforation. No lithic or organic with abrasive drill bits (Axial with parallel sides tip); at least 2 drill bits (2 distinct peaks on the diagram of holes diameters) in majority unipolar methods and after reamed; no use of hand perforator and bow drill, probably use of hand drill (circular outline, concentric and parallel striations);

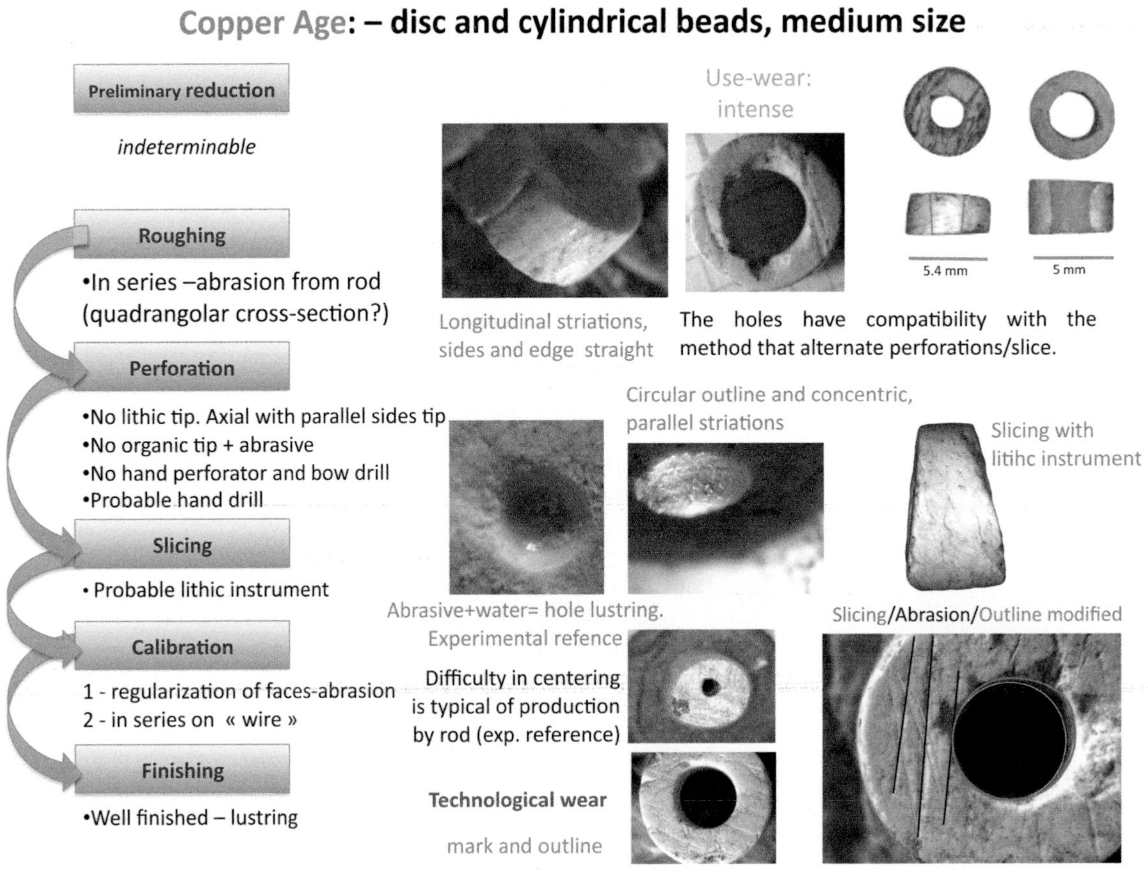

FIGURE 3. MANUFACTURE SEQUENCE OF COPPER AGE, DISC AND CYLINDRICAL BEADS-SMALL SIZE.

3. Slicing. The geometry (hole and sides) has compatibility with the method that alternate perforations and slice actions with lithic instrument.
4. Calibration. First a regularization of the faces by abrasion; after in series on 'wire'
5. Finishing. Well finished by lustring

Use wear: intense use wear (modifications of the perforation's outline).

Copper age: long beads (figure 4)

Archaeological context: grave 37-Remedello Sotto

Raw material: 1 carbonate

Morphometry: not very regular

Manufacture sequences:

1. Roughing. Individual by abrasion, first by transversal action (polygonal outline) and after by longitudinal action.
2. Perforation. Compatibility with metal tip without abrasive (perfect outline); bipolar method not reamed; no use of hand perforator and bow drill, probably use of hand drill (circular outline)
3. Finishing. Well polished

Use wear: light use wear (modifications of the perforation's outline).

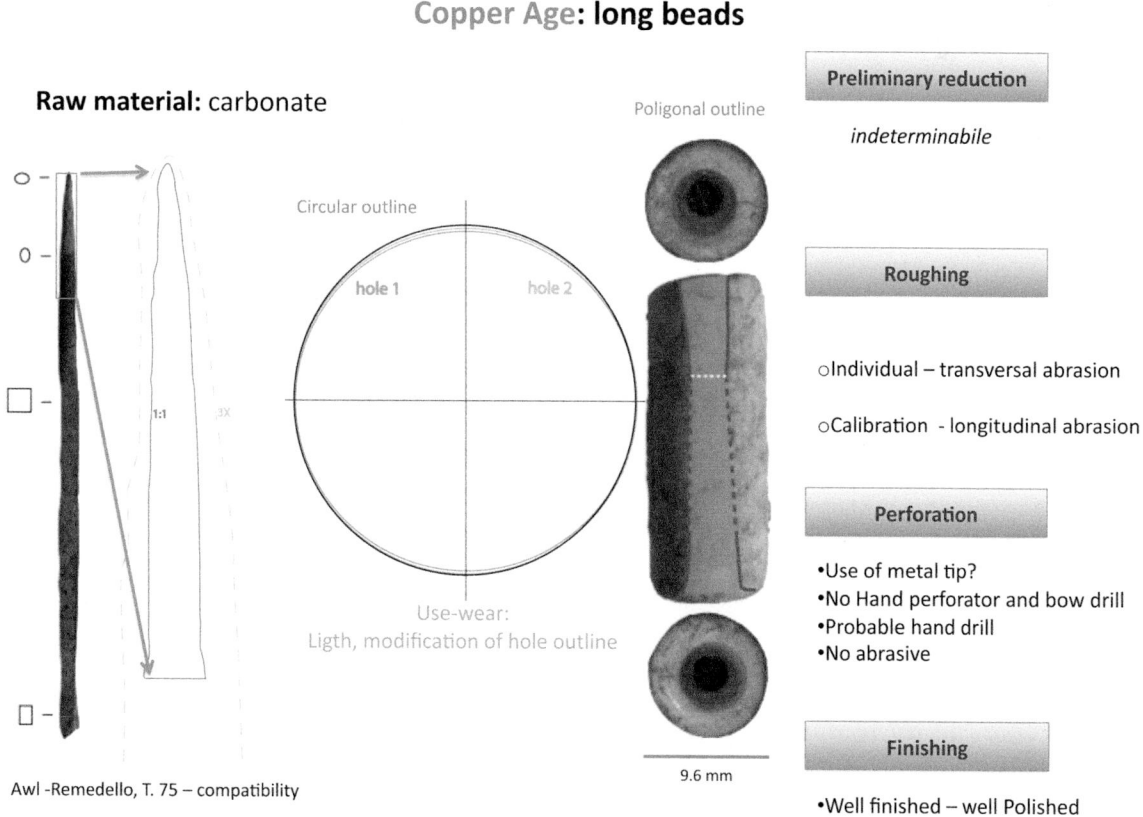

FIGURE 4. MANUFACTURE SEQUENCE OF COPPER AGE, LONG BEADS.

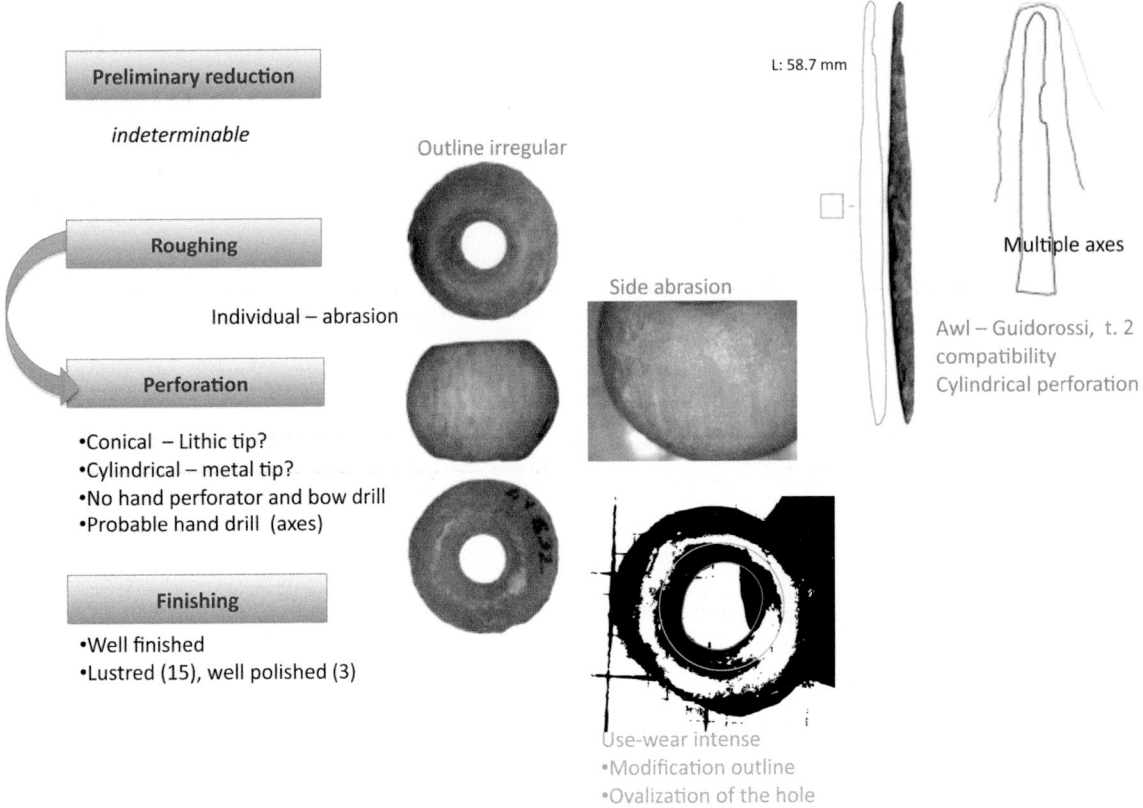

FIGURE 5. MANUFACTURE SEQUENCE OF BELL BEAKER CULTURE, BICONICAL AND GLOBULAR BEADS.

Copper age: biconical, globular, medium size (figure 5)

Archaeological context: grave 3-via Guidorossi

Raw material: 17 carbonate

Morphometry: not very standardized (14 biconical, 1 globular)

Manufacture sequences:

1. Roughing. Individual by abrasion
2. Perforation. 2 drill bits: lithic with hand drill (multiple axis) and bipolar method; a cylindrical, thin, unipolar perforation compatible with a metal tip; all perforations are not reamed; no use of hand perforator and bowdrill
3. Finishing. Lustred and well polished.

Use wear: intense use wear (ovalization of the perforation's outline).

Early Bronze age: archaeological data: disc, cylindrical, medium size (figure 6)

Archaeological context: grave 49-Arano

Raw material: 12 carbonate (calcite)

Early Bronze age: **discoids, cylindrical, medium size**

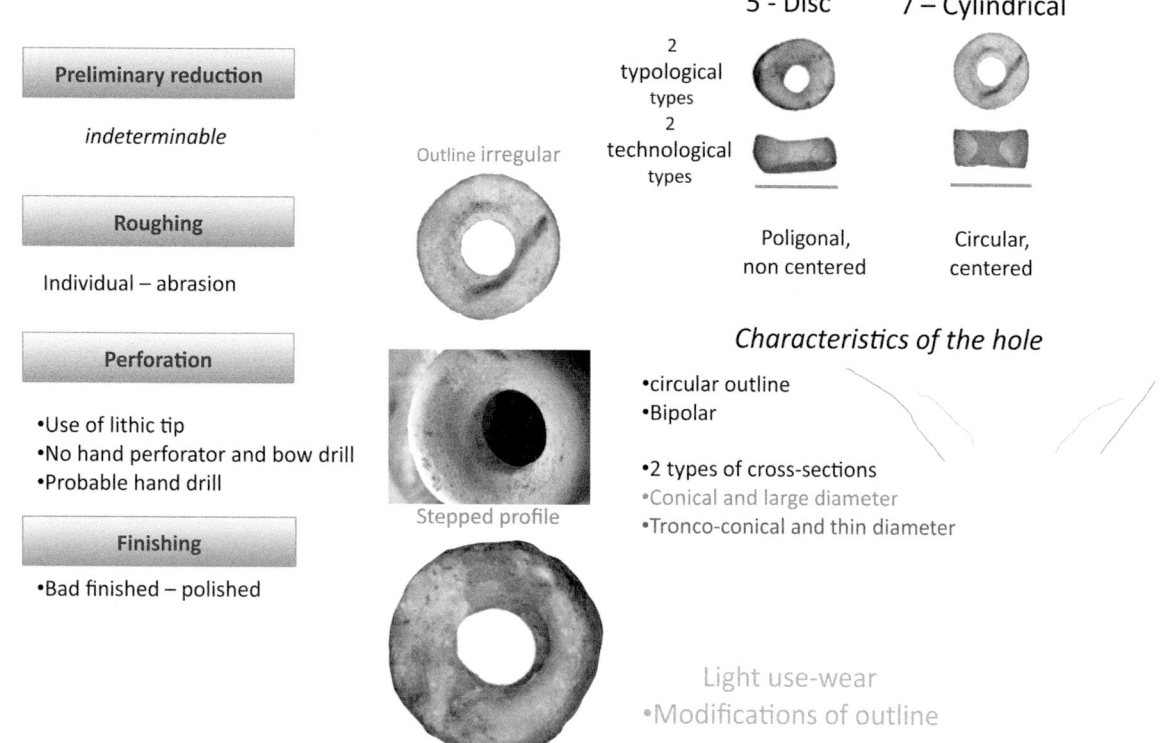

FIGURE 6. MANUFACTURE SEQUENCE OF EARLY BRONZE AGE, DISCOIDS, CYLINDRICAL BEADS-MEDIUM SIZE.

Morphometry: not standardized (5 discoid, 7 cylindrical). In this case, the typological types seem to be technological types.

Manufacture sequences:

1. Roughing. Individual by abrasion, to obtain 2 typological types: discoids
2. Perforation. 2 lithic drill bits (conical cross-section and large diameter and tronco-conical and thin diameter); bipolar method, not reamed; no use of hand perforator and bow drill.
3. Finishing. Poorly finished by simple polishing.

Use wear: very light or no use wear (modifications of the perforation's outline).

Discussion

The techno-functional study has highlighted several key aspects for the understanding of past societies from the point of view of their technology.

The first important aspect concerns raw material. As regards the studied sites (literature review and direct observation), there seems to be a clear choice in all sites belonging to the very similar raw materials, mainly, to the group of carbonate (calcite, marble, etc.) with very light colours that can be between the light and the white light yellow/gray.

In fact, the studied assemblages are very homogeneous. They are quite common in the geographical areas where the sites are located. So in the absence of quantitative compositional analysis, we cannot make an assumption about networks and methods of supply. While the soapstone is very easy to work, the carbonates are medium hardness materials. In the case of calcite, it tends to flake and is quite well suited to be polished and *lustrated* (in particular marble). Recently, it has been suggested that the raw material used for beads of Scalucce of Molina is not carbonate, but enstatite and very probably of synthetic origin (for details: Viola 2016, section 6.2). The exact determination requires appropriate quantitative analytical investigations, but if it were confirmed, it would provide new data on the good level of specialization in these processes during the Copper Age.

In any case, the homogeneity of the *corpus* may be the result of one or more factors still interacting as: ease of acquisition of raw material; certain technical criteria such as the relative ease of machining; but also less practical and more aesthetical criteria.

The second key aspect concerns craft specialization. During the Copper Age, in particular on smaller objects, we have good morphological standardization of supports (as tools of the drill) as well as a choice between different types (sometimes, perhaps, in combined use with loose abrasive) according to the objectives (for example the 2 clear peaks in the diameters of the perforations on Scalucce elements). The use of unipolar drilling seems usually prevalent but, in some cases, we also recognized complex methods of drilling such as multistage, or reaming, or the use of tips produced with complex processes (metal tips also for long perforations). The presence of metallic points during the Copper Age is quite probable (Viola 2016, ch. 6.4). For smaller elements (Scalucce di Molina site), it has been recognized as a complex mode of shaping with an in series method starting from a few rod, then processed through drilling and subdivision of sequences cycles.

Concerning the evaluation of the production volumes, it is not possible to express a precise estimation. Nevertheless, although the method of manufacturing in series is certainly much more difficult because it requires a good skills (in particular to produce very regular elements), it is also faster than the individual production because there is less raw material to be removed and drilled. So, in series method is suitable for large production of objects typical of specialized contexts. A very significant aspect in the production of small parts lies in the solutions adopted for overcoming technical difficulties imposed by the small size (for example, difficulty in avoiding erosion and risks of fracture during drilling). Those solutions are technological indicators of specialized craftsmanship. The degree of finish seems to be much more accurate during the Copper Age with extensive use of fine grinding or polishing of surfaces.

Overall, Early Bronze products are much less standardized and poorly finished, with less smooth surfaces and coarse perforations produced with faster methods (bow drill). Even when a more polished product is intentionally produced (short cylindrical beads of Arano site), it is still made with a lithic V profile tip without reaming. The drilling method selected in the latter site is clearly a bipolar method, perhaps to better manage the risks of typical calcite fractures.

The most interesting aspect concerns the various manufacturing sequences identified. In all cases, the initial stage of preliminary reduction is not observable since all objects are finished. The final finishing phase avoids the observational capabilities of the method. The steps that have been better recognized and offers the most information potential are the full production phases (shaping and drilling) in their various aspects (both morphological and morphometric).

In technological terms, the first techno-cultural variable lies in the choice of the drilling mechanism. In all cases, it never used the perforator hand. Indeed, this would have produced an irregular morphology of the holes. For the Copper Age, a generalized use of hand drill seems more likely. The cylindrical long bead of Remedello Sotto is the only object that can be excluded with certainty since it shows the use of the drill string. For the remaining elements of the necropolis, the morphologies are influenced by the deformations due to the use. However, for all other objects (thick or thin) from the Copper Age, it seems more likely that the hand drill was used and we can exclude the use of fast alternating movement drills or 'not rotary movement' through hand perforator.

For the Early Bronze Age site of Arano, depending on morphological type, it seems to have used a different mechanism: the cylindrical type with fine hole is most likely due to a slow movement while the discoid type to more flared hole is due to a fast moving product, perhaps with a bow drill.

Finally, the elements of Via Guidorossi, dated from the Copper Age before the Bell-Beaker culture, show some typical characteristics of the other two sites of Copper age (for example: heavy wear, be well refined, the use of different tips) but also a low morphological standardization. The information in some technological characteristics dimensions (for example the holes produced with bipolar method) may derive from the thickness of the support rather than to cultural reasons (most advanced stage of the Copper Age?).

In terms of technological traditions, the Copper Age shows:

- Selection of a specific raw material (homogeneous in nature and coloring).
- Objects smaller and with much standardized morphological relations.
- The sequence provides individual production of elements of larger size and production in series for smaller elements. These are produced with a method in series from rod.
- The drilling is done with a piercing mechanism to slow movement, most likely a hand drill employed with unipolar method. In this case, there is also no lack methods which deviate from the simple need to perform a perforation but require more complicated artisan knowledge as multistage or reaming. To do this, several tools have been used: different types of drills and abrasive.
- Finishing phase shows an aesthetic research towards better finished objects and tends to obtaining shiny surfaces.

We must add that if it was confirmed the determination of the synthetic origin for the raw material of the Scalucce di Molina site would be a documented case of highly specialized production.

The Bronze Age shows instead:

- Objects of larger size and less standardized forms.
- The sequence is divided into individual productions but differentiated in two technology types.
- The drilling phase occurs in two ways depending on the type: through the use of a perforating mechanism in fast motion (bow drill) and one in the slow movement (hand drill) with bipolar method. The mechanisms are armed only with lithic tips.
- The finishing is much less cared.

They are all elements in discontinuity with the previous period, except for the choice of raw material. Indeed, we see continuation of the selection of a specific stone (generally homogeneous in nature and light coloring).

Finally, aspects of the use wear of the beads. The functional study has shown that objects dated from the Copper Age are largely used and, in some cases, show obvious modifications, in particular on the edge and the contour of the perforations. To sum up, Copper Age jewelleries are worn objects (even intensively) and should reflect the finery of everyday life.

We have a completely different situation in Arano (Bronze Age site) where use wear is generally absent. It is reasonable to assume that the phenomenon may be related to a few changes in Bronze Age rituals: the studied objects could be jewellery made specifically for the burial and reflect more the codes of the rite than those of everyday life.

The context data do not allow us to issue any kind of attribution of specific jewellery to a genre, to an age group or the composite jewellery. The only certainty is that the ornamental elements are present in both the male and female burials.

Conclusion

The formal choices of themes, of manufacturing techniques and sequences, and the production contexts are strongly intertwined with each other and actively interact with the social context. Particularly in the case of stone beads, the *parure* is a concrete form of perpetuation and dissemination of social identity through the cultural groups in different chronological phases (Bains 2013).

The jewelry as a cultural product is the product accession to a set of specific rules of each human group. If we consider culture as part of the process by which societies adapted themselves to reality, then the ornamentation study is a way to deal with the daily life of the past (Viola 2016). Depending on demand and on the available raw material form, there are different ways of working that can be recognized on the finished objects: the physical plot is also a network of gestures, concepts, values and symbols.

Normally, in technological studies, the analysis of archaeological materials from production contexts (*ateliers*) is preferred, and their observation is made with high-magnification observation instruments (scanning electron microscope SEM). In the case in question, well finished objects have been studied, which could not be moved from the places of custody.

Therefore, a specific techno-functional analysis protocol for technological interpretation of some selected materials has been created, which could cope with such limitations. This protocol is based on an important bibliographic reference corpus (archaeological, ethnographic, ethno-anthropological), and led to the construction of an experimental research project, involving some aspects not investigated in the archaeological field, as yet

- Subdivision techniques
- Serial shaping modality
- Microperforations with different tips
- Types of reaming methods
- Tools and their wear

From a methodological standpoint, the proposed approach (ok?) has also attained the practical result of successfully testing a method of observation and recording based on portable instrumentation.

The archaeological materials chosen for the study come from recent excavations and are therefore unpublished to the scientific community (necropolis of Arano-Verona, Grave 3 via Guidorosssi-Parma), or belong to cultural contexts of considerable importance in the history of research, but have never received a complete publication (necropolis of Remedello-Brescia, necropolis of Scalucce-Verona).

The results of the techno-functional study have lived up to expectations. In fact, not only, for each site, have some important aspects of the manufacture sequencies (methods of shaping, of drilling and some used tools) been identified, but in many cases it was possible to gather the almost complete manufacturing sequences, and to determine whether each object was brought or not.

In the field of techno-economic interpretations, the study has confirmed a basic assumption of the northern Italian reconstructive framework: during the Copper age there the development of a specific specialized craft produces standardized and well finished products (see in Cocchi Genick 2011). However, also opposite results were achieved, such as the fact that in the Copper Age, in burials there was no use of specifically created objects for the deceased, but rather belonging to the sphere of everyday life. We can say this because these objects show wear to different degrees.

Some continuity between the Copper Age and the Bronze Age are nevertheless highlighted, such as the taste for certain types of composite objects, some raw materials and clear colors.

Also evident are at least two very significant discontinuities between the Copper Age and the Bronze Age: the jewels are produced in different ways, and the most recent objects were not used. In particular, these two discontinuities can be interpreted as a conflict between an ancient technological, tradition (Copper Age) with a new tradition (Early Bronze Age). The latter leads to much less finished productions and especially in objects produced specifically for burial. This, probably, to meet a need linked to a new funeral rite.

Foreword for reading the manufacture sequences: when the clear sequence of the gesture is identified an arrow (figures) or a number (text) links the phases to indicate sure and direct relation between them.

Acknowledgements

The session organizers who have accepted our communication, to Grant Augustin Lombard, SPHN de Genève for the economic helps, to Professor C. Perles for the suggestion that the raw material of Scalucce di Molina's beads are enstatite.

References

Bains, R., 2013 – *The social significance of Neolithic stone bead technologies at Catalhöyük*. London: University of London – PhD Thesis, 304 p.

Bernabo' Brea, M.; Miari, M., 2013 – Oltre il grande fiume: le necropoli dell'età del Rame in Emilia e Romagna. In De Marinis, R. C.; ed. lit. – *L'età del Rame. La pianura Padana e le Alpi al tempo di Ötzi*. Roccafranca: Compagnia della Stampa Massetti Rodella editori, p. 353-374.

Calley, S.; Grace, R., 1988 – Technology and function of microbores from Kumartepe (Turkey) In Beyries, S., ed. lit. – *Industries lithiques: tracéologie et technologie*. Oxford: B.A.R., p. 69-81. (BAR International Series, 441).

Chelidonio, G., 1988 – Experiments on boring-drilling technology: wear changes in tool shape and micro-wear. In Beyries, S., ed. lit. – *Industries lithiques: tracéologie et technologie*. Oxford: B.A.R., p. 285-308. (BAR International Series, 441).

Cocchi Genick, D., 2011 – Problematiche e porspettive della ricerca sull'età del Rame in Italia in ricordo di Gianni Balio Modesti. In Cocchi Genick, D., ed. lit – *L'età del Rame in Italia*. Firenze: Istituto Italiano di Preistoria e Protostoria, p. 13-21.

Coskunsu, G., 2008 – Hole-making tools of Mezraa Teleilat with special attention to micro-borers and cylindrical polished drills and bead production. *Neo-Lithics*. Berlin. 8: 1, p. 25-36.

Cornaggia Castiglioni, O., 1971 – La cultura di Remedello. Problematica ed ergologia di una facies dell'Eneolitico padano. *Memorie della Società Italiana di Scienze naturali e del Museo Civico di Storia Naturale di Milano*. Milano. 20:1, p. 1-79.

de Marinis, R., 2013 – La necropoli di Remedello Sotto e l'età del Rame nella pianura a nord del Po. In de Marinis, R. C.; ed. lit. – *L'età del Rame. La pianura Padana e le Alpi al tempo di Ötzi*. Roccafranca: Compagnia della Stampa Massetti Rodella editori, p. 301-351.

de Marinis, R.; Valzolgher, E., 2013 – Riti funerari dell'antica età del Bronzo in area padana. In de Marinis, R. C.; ed. lit. – *L'età del Rame. La pianura Padana e le Alpi al tempo di Ötzi*. Roccafranca: Compagnia della Stampa Massetti Rodella editori, p. 545-559.

Gurova, M.; Anastassova, E.; Bonsall, C.; Bradley, B.; Cura, P., 2014 – Experimental approach to prehistoric drilling and bead manufacturing. In Cura, S. (*et. al.*), eds. lits.- *Technology and Experimentation in archaeology*. Oxford: B.A.R., p. 47-55 (BAR International series, 2657:10).

Gurova, M.; Bonsall, C.; Bradley, B.; Anastassova, E., 2013 – Approaching prehistoric skills: experimental drilling in the context of bead manufacturing. *Bulgarian e-journal of Archaeology*. Sofia. 3, p. 201-221.

Farioli, E., 2015 – *Museo Chierici*. Reggio Emilia: Palazzo dei Musei Civici di Reggio Emilia, 84 p.

Valzolgher, E.; Lincetto, S., 2001-2002 – La necropoli eneolitica di Scalucce di Molina. Gli scavi De Stefani del 1883. *Annuario Storico della Valpolicella*. Verona. 18, p. 159-206.

Viola, S., 2016 – *Significato sociale della parure in pietra tra l'età del Rame e il Bronzo Antico dell'Italia settentrionale. Un approccio tecno-funzionale attraverso la sperimentazione archeologica*. Geneva: University of Geneva – PhD Thesis, 540 p.